Ethical and Philosophical Consideration of the Dual-Use Dilemma in the Biological Sciences

Seumas Miller • Michael J. Selgelid

Ethical and Philosophical Consideration of the Dual-Use Dilemma in the Biological Sciences

Springer

Seumas Miller
Centre for Applied Philosophy and Public Ethics
Charles Sturt University
The Australian National University
Canberra ACT
Australia

Michael J. Selgelid
Centre for Applied Philosophy and Public Ethics
Menzies Centre for Health Policy, and
National Centre for Biosecurity
The Australian National University
Canberra ACT
Australia

ISBN 978-1-4020-8311-2 e-ISBN 978-1-4020-8312-9

Library of Congress Control Number: 2008928679

Printed on acid-free paper

9 8 7 6 5 4 3 2 1

springer.com

Acknowledgements

This book was originally produced as a report for the National Security Science and Technology Unit, Department of the Prime Minister and Cabinet, Australian Government, as part of a Centre for Applied Philosophy and Public Ethics (CAPPE) consultation in 2006. It includes contributions from Antony Della-Porta (Biosecurity and Biocontainment International Consultants Pty Ltd); Christian Enemark (University of Sydney); Peter Kerr (Commonwealth Scientific and Industrial Research Organisation); and Ian Ramshaw (The Australian National University). It was also improved thanks to interaction with, and feedback from, Rob Floyd (Department of the Prime Minister and Cabinet) and Tony Willis (Department of Foreign Affairs and Trade). The Centre for Applied Philosophy and Public Ethics is an Australian Research Council funded Special Research Centre (Australian National University, Charles Sturt University, University of Melbourne). Miller and Selgelid are equal co-authors. This report was previously published in Science and Engineering Ethics (2007) 13: 523–580.

Contents

Acknowledgements v

Executive Summary 1

1 Introduction 9

2 Experiments of Concern 17

3 The Ethics of Dual-Use Research 31

4 Dissemination of Dual-Use Research Results 39

5 Options for the Regulation of Dual-Use Experiments and Information 49

Bibliography 65

Index 73

Executive Summary

What Is the Dual-Use Dilemma?

The so-called "dual-use dilemma" arises in the context of research in the biological and other sciences as a consequence of the fact that one and the same piece of scientific research sometimes has the potential to be used for harm as well as for good.

A dual-use dilemma is an *ethical* dilemma, and an ethical dilemma for the *researcher* (and for those who have the power or authority to assist or impede the researcher's work, e.g., governments). It is an *ethical* dilemma since it is about promoting good in the context of the potential for also causing harm, e.g., the promotion of health in the context of providing the wherewithal for the killing of innocents. It is an ethical dilemma *for the researcher* not because he or she is aiming at anything other than a good outcome; typically, the researcher intends no harm, but only good. Rather, the dilemma arises for the researcher because of the potential actions of *others*. Malevolent non-researchers might steal dangerous biological agents produced by the researcher; alternatively, *other* researchers—or at least their governments or leadership—might use the results of the original researcher's work for malevolent purposes. The malevolent purposes in question include bioterrorism, biowarfare and blackmail for financial gain.

In the aftermath of the 11th September 2001 attacks and the subsequent anthrax letters episode in the US, bioterrorism is widely considered to be a real threat, especially to populations in western countries. Moreover, it is seen as a more likely threat from non-state terrorist groups than, say, nuclear weapons of mass destruction, given the availability of the technical knowledge necessary to produce the relevant biological agents and the feasibility of weaponisation. (In this report we assume terrorist acts could be performed by state actors as well as non-state actors. This is consistent with the definition given in the Commonwealth of Australia Criminal Code 1995.)

Aims and Scope of This Report

In general terms, the aims and scope of this report are to provide a reasonably comprehensive array of (possibly competing) answers to the following questions:

S. Miller, M.J. Selgelid, *Ethical and Philosophical Consideration of the Dual-Use Dilemma in the Biological Sciences,*

(A) Morally Impermissible Research:

- What, if any, research in the biological sciences that does *not* give rise to a dual-use dilemma is morally impermissible, e.g., research undertaken for purely offensive military purposes?
- What is the dual-use dilemma in the biological sciences, and in what categories of research does it arise, e.g., experimental research undertaken to assist in the combating of mice plagues that might in fact result in the development of a more virulent form of smallpox?
- What are the moral and other considerations in play in these various categories of research that give rise to dual-use dilemmas, e.g., potential to save human life versus potential to destroy human life?
- In light of these considerations what, if any, research in the biological sciences that gives rise to dual-use dilemmas is morally impermissible?
- Who is to decide what research, if any, in the biological sciences is morally impermissible, e.g., biosecurity committees?

(B) Physical and Regulatory Conditions under which (Permissible) Experiments of Concern Ought to be Undertaken:

- In relation to the various categories of *prima facie* permissible research that, nevertheless, give rise to dual-use dilemmas, what are the safety and security—and associated regulatory—conditions under which this research ought to be undertaken, e.g., background checks and security clearance for research personnel?

(C) Dissemination:

- What are the moral and other considerations in play in relation to the ownership rights (intellectual property) of permissible, safe and secure research in the biological sciences that, nevertheless, gives rise to dual-use dilemmas?
- What are the moral and other considerations in play in relation to the dissemination of findings from permissible, safe and secure research in the biological sciences that, nevertheless, gives rise to dual-use dilemmas?
- In relation to permissible, safe and secure research in the biological sciences that, nevertheless, gives rise to dual-use dilemmas what, if any, restrictions ought to be placed on dissemination of research findings?
- In relation to permissible, safe and secure research in the biological sciences that, nevertheless, gives rise to dual-use dilemmas who ought to decide what, if any, research findings ought not to be disseminated or ought to have restrictions placed on their dissemination?

Experiments of Concern

In relation to the dual-use dilemma in the biological sciences, the approach of the US National Research Council (NRC) in its influential 2004 report, *Biotechnology Research in an Age of Terrorism*, is to map the range of dual-use dilemmas by identifying and taxonomising a set of salient "experiments of concern". We accept this approach in the context of our attempt to isolate the morally permissible from the morally impermissible in relation to dual-use research in the biological sciences.

According to the NRC report "experiments of concern" are those that would:

1. Demonstrate how to render a vaccine ineffective
2. Confer resistance to therapeutically useful antibiotics or antiviral agents
3. Enhance the virulence of a pathogen or render a non-pathogen virulent
4. Increase the transmissibility of a pathogen
5. Alter the host range of a pathogen
6. Enable the evasion of diagnosis and/or detection by established methods or
7. Enable the weaponization of a biological agent or toxin [102].

Other possible categories are:

8. Genetic sequencing of pathogens
9. Synthesis of pathogenic micro-organisms
10. Any experiment with *variola* virus (smallpox) or
11. Attempts to recover/revive past pathogens.

Ethical Analysis of Dual-Use Research

Fine-grained ethical analyses of dual-use research in the biological sciences would seek to *quantify* actual and potential benefits and burdens, and actual and potential recipients/bearers of these benefits and burdens. These analyses would also identify a range of salient policy options. Each option would embody a set of trade-offs between present and future benefits and burdens, and recipients and bearers thereof. The construction of these options and the process of selection between them would consist in large part in the application of various ethical principles, including human rights principles—e.g., right to life, freedom of inquiry, and free speech—and principles of utility and justice.

We are not in a position to provide such fine-grained ethical analysis but will rather focus (somewhat simplistically) on a single ethical consideration—namely, human health (including human life)—that gives rise to the dilemma, and we will do so without exploring questions of which human populations or how many individual humans have benefited/been burdened or are likely to benefit/be burdened, and so on. Viewed from this perspective the dual-use dilemma concerns human health (as a simple, unquantified human good), and the dilemma consists in the fact that research undertaken to promote human health might instead be used to destroy human health. As such, the dilemma gives rise to questions of security. What are reasonable and ethically justified forms and degrees of security in this context?

The above mode of analysis of the dual-use dilemma consisting of the quantification of harms and benefits, the identification of salient options, and the selection of an option on the basis of ethical principles is reasonable as far as it goes. However, a more dynamic, indeed creative, mode of analysis is called for.

In the first place, options are not static because well-intentioned scientists, malevolent actors and security personnel are *responsive* to the problems that they confront, including the problems posed by other actors. The response of scientists to a pathogen with enhanced virulence might be the development of a new vaccine. The response of security personnel to a new bioterrorist threat might be an enhanced regulatory system. Accordingly, the mode of analysis of the dual-use dilemma must be dynamic in character.

In the second place, ethical dilemmas are not necessarily—or even typically—to be resolved by careful calibration of the differential ethical weight that attaches to the options provided for in the dilemma. Rather the dilemma must, if possible, be resolved by *designing* a new third or fourth option, i.e. by bypassing the dilemma. Consider the question of whether to disseminate dual-use research findings or not disseminate them: academic freedom versus security. Perhaps the solution is to find a third option, such as to disseminate them in a manner that will not enable the experiments in question to be replicated. This mode of analysis is *creative*. It lets us have our cake and eat it; it squares the ethical circle. Let us refer to this kind of ethical analysis as *designing-in ethics*.

In relation to the issue of the moral permissibility of dual-use research, there appear to be three separate (kinds of) ethical questions. First is the ethical question whether or not a putative biological agent to be researched ought in fact to be eliminated (or, if already eliminated, not retrieved). Here the *possibility* of research is removed; no possibility of research because no biological agent to be researched. We have in mind the case of smallpox and the arguments in favour or against the elimination of all samples of smallpox. (We note that in the new world of genetic sequencing and synthetic genomics it may not matter so much whether the wild-type virus actually exists; since it might be recreated from the sequence alone.) Second are the ethical questions arising from dual-use research in relation to a biological agent whose present and/or future existence is taken as a given (i.e., there is no intention to eliminate or not retrieve or not bring into existence the biological agent in question). For example, research to determine whether or not avian influenza could trigger a human pandemic might lead to the creation of dangerous new strains that could be used by terrorists. Such research might include work intentionally undertaken to create novel pathogens or synthesising existing ones, albeit work whose ultimate purpose was to develop, say, a vaccine against these pathogens. Third is the ethical question of whether to undertake dual-use research for the purpose of protection against weaponised pathogens, e.g., research into the aerosolisation of pathogens.

The smallpox debate is essentially about balancing the desirability of retaining the virus for prospective research purposes against the uncertain risks associated with not destroying it. In relation to the second questions there is a complex mix of scientific, security and public health questions, as well as ethical ones. Moreover, because the actors involved in dual-use dilemmas are responsive to problems and to one another's actions, probability judgments regarding risks need to take this into account. One way to do so is to analyse, for example, a security risk from bioterrorists in part in terms of a complex set of variables including the ability, opportunity and motivation of the bioterrorists, the likely intelligence possessed by, and the likely assessment made by, the terrorists, the capacity to respond to specific forms of bioterrorist attack, the likely movements of innocent third parties at risk from specific security responses, the relevant moral principles, the rights and duties of the various actors involved, and so on. However, what might be crucial here is the capacity to generate a creative response to the security problem thus analysed. Perhaps a focus on reducing the opportunities available to bioterrorists by establishing a licensing system for laboratories using dual-use technologies is a case in point.

In relation to the third question, the issue is whether or not in practice the weaponisation of pathogens for protective purposes can be distinguished from the weaponisation of pathogens for offensive purposes. Presumably, if these two conceptually distinct activities are to be distinguished in practice then this is because there are *verifiable* differences regarding: (i) intention or purpose; and/or (ii) physical properties of the weaponised pathogen.

Options for the Regulation of Dual-Use Experiments and Information

In what follows we provide a set of options regarding the imposition of limits on dual-use experiments and the dissemination of potentially dangerous information concerning dual-use discoveries. The five options presented range from the least intrusive/restrictive to the most intrusive/restrictive, and the advantages and disadvantages of each option are discussed in turn. Each option is complex in that it consists of six sub-options each of which pertains to one of the following six categories of decision.

(1) Permissible or Impermissible Research
Who is to be the decision-maker in relation to determining whether or not an instance of one of the eleven identified types of experiment of concern is permissible or impermissible? The candidates for decision-maker are: the individual researcher; the specific institution hosting the research project in question, i.e. a university, corporation or government research centre; an independent authority; the government. In the case of the university, the decision-maker would presumably be a collegial body comprised of relevant scientists (at least).

Freedom of inquiry is a human right that finds institutional expression in universities in the form of academic freedom. In the context of a liberal democracy there is a presumption against governmental restriction of human rights, including in the name of protecting other human rights. Moreover, arguably progress in science is importantly dependent on academic freedom. The question of whether research is morally permissible or impermissible is an extremely difficult issue, and it is by no means obvious who the ultimate decision-maker ought to be.

(2) Mandatory Physical Safety and Security Regulation
Should there be regulations regarding the safety and security of the storage, transport and physical accessibility of samples of pathogens, equipment, laboratories, and so on? The answer is presumably in the affirmative.

In theory, the specific content of these regulations might be determined either by a government agency, an independent authority, a professional association of scientists or the specific institution hosting the research in question. However, governments bear the ultimate institutional and moral responsibility for the safety and security of their citizens, including the researchers themselves, in so far as that safety and security is a matter of the physical conditions under which potentially harmful (albeit permissible) research is to be undertaken and the physical elements thereof stored, transported etc. Accordingly, the government would at least need to be able to satisfy itself that the regulatory system, including the regulations and their enforcement mechanisms, governing the physical safety and security of dual-use experimentation are adequate.

The application of many of these regulations could be undertaken by, for example, biosafety committees operating at the institutional level, e.g., a university-based biosafety committee. However, these committees would need in turn to be accountable to government (perhaps via an independent authority).

(3) Licensing of Dual-Use Technologies/Techniques
Should there be mandatory licensing of dual-use technologies/techniques/pathogen samples? Only certain laboratories in the public sector and the private sector might be licensed to engage in research involving the use of certain dual-use technologies.

The establishment of a licensing authority would be a significant addition to the mechanisms available to contain the dangers associated with the dual-use dilemma. However, it raises a number of important questions. One set of question concerns the criteria that the licensing authority would deploy in its licensing processes. Is there a presumption in favour of granting a licence; the criteria having been framed for the sole purpose of eliminating licence applicants that are manifestly unable to provide a safe and secure research environment? Are the criteria to be used for issuing of licences objective and publicly available? Another set of questions pertains to the status and make-up of the licensing authority. Is it, for example, independent of government in the sense that its decisions are binding and not able to be overridden by government?

These questions are important in the context of the concerns one might have in relation to government interference with freedom of intellectual inquiry (who decides what is permissible research?) and freedom of speech/dissemination of research findings (who decides what research findings can be disseminated and to whom?). The point is that a licensing authority could be given powers that would in effect override human rights to freedom of intellectual inquiry and freedom of speech.

(4) Mandatory Education and Training
Given the potential harms arising from the eleven identified types of experiments of concern it is clear that some process of education and/or training for relevant

researchers and other personnel is called for. There are questions about the precise content of such education and training. However, at the very least those working in laboratories need to receive training in relation to the safety and security of the storage, transport and physical accessibility of pathogens, equipment, laboratories, and so on. In addition, there is a need to ensure that editors and others responsible for the dissemination of potentially harmful information are aware of the issues in relation to dual-use research findings. In short, some forms of mandatory education and/or training are justified. What the precise content of such education/training programs ought to be and who ought to be responsible for their provision remain open questions. However, it is an institutional and moral responsibility of government to ensure that minimal training/education programs in relation to potentially harmful dual-use research and dissemination of dual-use research findings are being provided (even if not by government itself).

(5) Mandatory Personnel Security Regulation

Physical safety and security of a research environment, including access by non-authorised persons, e.g., potential thieves, is one thing; however, personnel security in relation to researchers, e.g., background checks, screening of researchers in relation to any history of mental illness, political affiliations with extremist groups etc., is quite another. Doubtless it is prudent, indeed it is a moral requirement, that access to virulent pathogens be disallowed to a researcher diagnosed as a psychopath or to a known member of a terrorist organisation. On the other hand, other things being equal, government officials prying into the lives of university students enrolled in degrees in the biological sciences is an unwanted intrusion on civil liberties. Here, as elsewhere, the devil is in the detail, and there is a need for specific policies to be framed in the light of a range of human rights and academic and scientific considerations as well as security concerns.

One way forward here might be to develop a system of security checks for personnel working in licensed laboratories but (absent special considerations) not for other research personnel.

(6) Censorship/Constraint of Dissemination

The question of whether research findings ought to be freely disseminated, censored or otherwise restricted in their dissemination is an extremely difficult issue and it is by no means obvious who the ultimate decision-maker ought to be. Freedom of speech and freedom of dissemination of knowledge are human rights that find institutional expression in universities in the form of academic freedom. In the context of a liberal democracy there is a presumption against governmental restriction of human rights, even in the name of protecting other human rights. Moreover, arguably progress in science is importantly dependent on academic freedom.

A relevant important distinction here is that made between 1st tier and 2nd tier dual-use research. First tier dual-use research findings might need to be disseminated in a manner that will not enable those being informed to replicate the experiments in question.

The five broad options and their sub-options can be represented in the form of a matrix, as shown in Table 1.

Table 1 Decision-making for dual-use dilemmas in the biological sciences

	Options				
Decisions	Option 1—the complete autonomy of the individual scientist	Option 2—institutional control	Option 3—institutional & governmental control	Option 4—an independent authority	Option 5—governmental control
Who are the decision-makers regarding im/permissible research?	Individual researcher	(i) Scientists in university (collegial) (ii) Corporation (iii) Govt. Res. Centre	(i) Scientists in university (collegial) (ii) Corporation (iii) Govt. Res. Centre	Independent authority	Government
Should compliance with physical safety & security regulation be mandatory?	No	Yes	Yes	Yes	Yes
Should dual-use technology be licensed?	No	No	Yes	Yes	Yes
Should education & training be mandatory?	No	No	Yes	Yes	Yes
Should personnel security regulation be mandatory?	No	No	Yes	Yes	Yes
Who are the decision-makers regarding censorship/constraint of material proposed for dissemination?	Individual editor	(i) Individual editor (ii) Corporation (iii) Govt. Res. Centre	(i) Individual editor (ii) Corporation (iii) Govt. Res. Centre	Independent authority	Government

NB: The decision-making in question pertains only to dual-use research in the biological sciences identified as potentially problematic by virtue of coming under one of the pre-established headings of Experiments of Concern

Chapter 1
Introduction

What Is the Dual-Use Dilemma?

The so-called "dual-use dilemma" arises in the context of research in the biological and other sciences as a consequence of the fact that one and the same piece of scientific research sometimes has the potential to be used for harm as well as for good. Consider as an example of this kind of dilemma recent research on the mousepox virus [26]. On the one hand, the research program on the mousepox virus should have been pursued since it may well have led to a genetically engineered sterility treatment that would have helped combat periodic plagues of mice in Australia. On the other hand, this research project should not have been pursued since it led to the creation of a highly virulent strain of mousepox and the possibility of the creation—by, say, a terrorist group contemplating a biological attack—of a highly virulent strain of smallpox resistant to available vaccines.

A dual-use dilemma is an ethical dilemma, and an ethical dilemma for the researcher as well as for those (e.g., governments) who have the power or authority to assist or impede the researcher's work. It is an ethical dilemma since it is about promoting good in the context of the potential for also causing harm, e.g., the promotion of health in the context of providing the wherewithal for the killing of innocents. It is an ethical dilemma for the researcher not because he or she is aiming at anything other than a good outcome; typically, the researcher intends no harm, but only good. Rather, the dilemma arises for the researcher because of the potential actions of others. Malevolent non-researchers might steal dangerous biological agents produced by the researcher; alternatively, other researchers—or at least their governments or leadership—might use the results of the original researcher's work for malevolent purposes. The malevolent purposes in question include bioterrorism, biowarfare and blackmail for financial gain.

In the recent and not so recent past, a number of governments have sought to develop weapons of mass destruction (WMDs), including biological weapons, and in some cases have actually used them, e.g., the use of mustard gas by the German and British armies in World War I (WW1), the dropping of atomic bombs on Hiroshima and Nagasaki by the US Air Force in World War II (WW2), the existence

S. Miller, M.J. Selgelid, *Ethical and Philosophical Consideration of the Dual-Use Dilemma in the Biological Sciences*,

of a large-scale biological weapons program in the Soviet Union from 1946 to 1992, and the use of chemical agents against the Kurds by Saddam Hussein's regime in 1988.

Moreover, there have been some high profile "defections" of scientists from western countries to authoritarian states with WMD programs. For example, Dr. Abdul Qadeer Khan joined and in large part established Pakistan's nuclear weapons program after working for Urenco in the Netherlands; and Frans van Anraat (also from the Netherlands) went to Iraq to assist Saddam Hussein's WMD program in producing mustard gas.

Further, there have been a number of acts, or attempted acts, of bioterrorism, notably by the Aum Shinrikyo in Japan (they attempted to acquire and use anthrax and botulinum toxin, and they successfully launched an attack with Sarin in a Tokyo Subway), Al-Qaeda (they attempted to acquire and use anthrax) and the so-called Amerithrax attacks (involving the actual use of anthrax).

In the aftermath of the 11th September 2001 attacks in the US, bioterrorism is widely considered to be a real threat, especially to populations in western countries. Moreover, it is seen as a greater threat from non-state terrorist groups than, say, nuclear WMDs, given the availability of the materials and technical knowledge necessary to produce the relevant biological agents and the feasibility of weaponisation. This is not to say that there are not obstacles for would-be bioterrorists, including the dangers to themselves in handling pathogens. But it is to say that there is a non-negligible bioterrorist threat, and it is likely to increase rather than decrease.

A small number of animal, plant and human pathogens are readily obtainable from nature, and bioterrorists with minimal microbiological training could use these to inflict causalities or economic damage. Techniques of genetic engineering have been available for some time to enhance the virulence, transmissibility and so on of naturally occurring pathogens. This gives rise to the possibility of terrorists getting their hands on pathogens with (say) enhanced virulence and for which there are no vaccines or treatments. Indeed, some of these enhancement techniques are such that bioterrorists with advanced microbiological training could themselves employ them.

Recent developments in synthetic genomics have exacerbated the problem even further. It is now possible to create pathogens *de novo*, i.e., to construct deadly viruses from scratch. Accordingly, in the not too distant future a would-be terrorist will no longer need to go to an inhospitable region to find a naturally occurring pathogen such as Ebola, or to steal a highly virulent and transmissible pathogen such as smallpox from one of a very small number of very secure laboratories, or even to employ standard recombinant DNA techniques to enhance the virulence and transmissibility of some more readily available pathogen. Rather he or she could buy a bench-top DNA synthesiser and potentially use it to assemble a specified genomic sequence of a highly virulent and transmissible pathogen from readily available raw materials.

Again, this is not to say that there are not obstacles to terrorists, including those mentioned above, as well as the current lack of know-how and technological capability regarding synthetic genomics amongst most cohorts of researchers and laboratory workers, and whatever safeguards exist now, e.g., the US Select Agent regulations, or can be put into place over the next few years.

In short, *some* research in the biological sciences has the potential for great harm, as well as great good; and, unfortunately, there are any number of malevolent individuals, political and religious groups and governments ready, willing and (increasingly) able to use this research to cause harm rather than to do good. This is the larger context in which the dual-use dilemma in the biological sciences arises.

The expression "dual-use dilemma" is in need of some conceptual unpacking. Here we need to introduce a number of sets of distinctions.

(1) In relation to the *purposes* (or ends) of the research, we can distinguish the following conceptual axes: (i) good/harmful; (ii) military/non-military; and (iii) within the category of military purposes, the sub-categories of offensive/protective. Consider the aerosolisation of a pathogen undertaken for a military purpose. The purpose in question might be offensive, e.g., biowarfare; but it might simply be protective, e.g., to understand the nature and dangers of such aerosolisation in order to prepare protections against an enemy known to be planning to deploy the aerosolised pathogen as a weapon.

    The categories good/harmful and military/non-military do not necessarily mirror one another. Some non-military purposes are, nevertheless, harmful, e.g., the supplier of a drug releasing a pathogen to make large numbers of people sick in order that the sick buy the drug against the pathogen and, thereby, increase the supplier's profits. And some military purposes might be good, e.g., the above-mentioned research on the aerosolisation of a pathogen undertaken for purely protective purposes in the context of a just war. The United States Project BioShield is an example of research aimed at providing "new tools to improve medical countermeasures protecting Americans against a chemical, biological, radiological or nuclear (CBRN) attack" [134]. However, some of the protective research would probably yield results that could assist in the development and delivery of biological weapons.

(2) Dual-use refers to two logically distinct "users" of the research: (i) those who initially undertake the research or use the research results for purposes intended by the original researchers (original/intended users); and (ii) those who use the research results for some purpose other than that intended by the original researchers (secondary users). For example, the above-mentioned research on the aerosolisation of a pathogen (conducted by original/intended users) might be used for offensive purposes by those fighting an unjust war (the secondary users).

(3) In relation to the term "use" we can distinguish: (i) actually or potentially used in accordance with the purpose for which it was designed (design-purpose); (ii) actually or potentially used for some purpose other than that for which it was specifically designed. Dual-use dilemmas can involve original researchers whose purpose is a design-purpose, e.g., demonstrate how to render a vaccine against a highly transmissible pathogen ineffective. This design-purpose can

itself be in the service of a benevolent purpose of the original researchers, e.g., the purpose of enhancing the effectiveness of the vaccine. Alternatively, the achievement of this design-purpose could be used for a malevolent non-design purpose by secondary researchers, e.g., to render the vaccine ineffective in the context of spreading the pathogen in question.

On the other hand, secondary users might build on the original research in such a way as to create, say, a new pathogen, e.g., a more virulent strain of smallpox as opposed to a more virulent strain of mousepox, in which case we might be inclined to say that they had a new *design*-purpose (albeit a malevolent one).

(4) In relation to the *outcomes* of the research, we can distinguish: (i) intended outcomes; (ii) unintended but foreseen outcomes; and (iii) unforeseen and perhaps unable to have been foreseen outcomes. An example of an unintended outcome is an outbreak of smallpox resulting from inadequate safety procedures in a laboratory setting. However, such accidents are not instances of the dual-use dilemma. For something to be an instance of a dual-use dilemma, both outcomes (the two horns of the dual-use dilemma) need to be (actually or potentially) intended (or at least foreseen) by someone; there needs to be two sets of (actual or potential) *users*. Naturally, an outcome might be unintended and unforeseen (even unforeseeable) by the original researcher but, nevertheless, intended by the secondary user. Thus, scientists who preserve a small number of smallpox samples for pure research purposes in the context of a policy of mandatory destruction of samples might not intend or foresee that they might be used for malevolent purposes by others, e.g., weaponised.

The dual-use dilemma is obviously a dilemma for researchers, viz. those researchers involved in biological research that has the potential to be misused by bioterrorists, criminal organisations and governments engaged in biowarfare. But it is also a dilemma for the private and public institutions, including universities, that fund or otherwise enable research to be undertaken. The dilemma is made more acute for university-based researchers and for universities, given their commitments to such values as academic freedom and the unfettered dissemination of research findings; and for private companies, given their commitment to free-enterprise. More generally, it is a dilemma for the individual communities for whose benefit or, indeed, to whose potential detriment, the research is being conducted, and for the national governments who bear the moral and legal responsibility of ensuring that the security of their citizens is provided for. Finally, in the context of an increasingly interdependent set of nation-states—the so-called global community—the dual-use dilemma has become a dilemma for international bodies such as the United Nations.

Biological Weapons Convention

Given the general threat to public health posed by transmissible pathogens, and given that biological agents can be used as WMDs in the hands of state actors, terrorist groups and criminal organisations, there is an imperative to strictly regulate the development, production, stockpiling, weaponisation and use of pathogens. At the international level, a key instrument in this regard is the Biological Weapons Convention (BWC)—more precisely, *Convention on the Prohibition of the Development, Production and Stockpiling of Bacteriological (Biological) and Toxin Weapons and on Their Destruction.* (Signed at London, Moscow and Washington on 10th April 1972. Entered into force on 26th March 1975. Depositories—UK, USA and Soviet governments.)

The general aim to which the BWC is directed is, "for the sake of mankind, to exclude completely the possibility of bacteriological (biological) agents and toxins being used as weapons. Convinced that such use would be repugnant to the conscience of mankind and that no effort should be spared to minimise this risk".

In accordance with Article 1 of the BWC,

> Each State Party to this Convention undertakes never in any circumstances to develop, produce, stockpile or otherwise acquire or retain:
>
> 1. Microbial or other biological agents, or toxins whatever their origin or method of production, of types and in quantities that have no justification for prophylactic, protective or other peaceful purposes
> 2. Weapons, equipment or means of delivery designed to use such agents or toxins for hostile purposes or in armed conflict

While the BWC is an important step in relation to its stated aims of prohibiting and eliminating the possibility of using biological agents as weapons, it has a number of possible loopholes and lacunae.

The BWC evidently has requirements regarding technology transfers from prohibited to non-prohibited purposes and vice-versa [70]. For example, technology transfers from non-prohibited purposes, e.g., prophylactic, to prohibited, e.g., military offensive, are prohibited under all circumstances.

However, the BWC does not make a formal distinction between civilian and military purposes. Indeed, in speaking of "protective purposes" (clause 1, above) the BWC seems to allow protective military purposes. This has the consequence that a technology transfer from civilian to military is allowable, if the latter purpose is protective and not offensive. But now an issue arises as to what counts as protective, as opposed to offensive. (See below for more on this issue.) Moreover, the BWC does not provide for any robust verification processes, e.g., unlike the Chemical Weapons Convention (CWC) there is no international organisation or national authority to verify compliance with the BWC.

Aims and Scope of This Report

In general terms, the aims and scope of this report are to provide a reasonably comprehensive array of (possibly competing) answers to the following questions (and to the further more specific questions to which these answers give rise).

Morally impermissible research:

- What, if any, research in the biological sciences that does not give rise to a dual-use dilemma is morally impermissible, e.g., research undertaken for purely offensive military purposes?
- What is the dual-use dilemma in the biological sciences, and in what categories of research does it arise, e.g., experimental research undertaken to assist in the combating of mice plagues that might in fact result in the development of a more virulent form of smallpox?
- What are the moral and other considerations in play in these various categories of research that give rise to dual-use dilemmas, e.g., potential to save human life versus potential to destroy human life?
- In the light of these considerations what, if any, research in the biological sciences that gives rise to dual-use dilemmas is morally impermissible?
- Who is to decide what research, if any, in the biological sciences is morally impermissible, e.g., biosecurity committees?

Physical and regulatory conditions under which (permissible) experiments of concern ought to be undertaken:

- In relation to the various categories of *prima facie* permissible research that, nevertheless, give rise to dual-use dilemmas, what are the safety and security— and associated regulatory—conditions under which this research ought to be undertaken, e.g., background checks and security clearance for research personnel?

Dissemination:

- What are the moral and other considerations in play in relation to the ownership rights (intellectual property) of permissible, safe and secure research in the biological sciences that, nevertheless, gives rise to dual-use dilemmas?
- What are the moral and other considerations in play in relation to the dissemination of findings from permissible, safe and secure research in the biological sciences that, nevertheless, gives rise to dual-use dilemmas?
- In relation to permissible, safe and secure research in the biological sciences that, nevertheless, gives rise to dual-use dilemmas what, if any, restrictions ought to be placed on dissemination of research findings?
- In relation to permissible, safe and secure research in the biological sciences that, nevertheless, gives rise to dual-use dilemmas who ought to decide what, if any, research findings ought not to be disseminated or ought to have restrictions placed on their dissemination?

Note that our primary concern in this report is with moral or ethical (we use the terms interchangeably) principles and values, as opposed to legal or, for that matter, regulatory rules. There is, of course, a close relationship between the moral and the legal. For instance, typically criminal laws, such as the laws against murder, assault and theft, 'track' or follow antecedent moral principles; there is a law against murder, for example, precisely because we regard murder as *morally* wrong. Nevertheless, the moral and the legal are conceptually distinct, and the distinction needs to be kept in mind in what follows. An important corollary of the existence of this moral/legal distinction is that it is not necessarily the case that every research practice rightly regarded as immoral or unethical should always be made unlawful.

Chapter 2
Experiments of Concern

Human knowledge and understanding of the natural world is, presumably, both desirable in itself and a means to the provision of other human goods, such as health and longevity. Moreover, human freedom, including freedom of intellectual inquiry, is agreed on all hands to be an intrinsic human good. Accordingly, there is a presumption in favour of allowing research in the biological sciences, as there is in other areas of human knowledge. In short, research in the biological sciences is morally permissible, absent special considerations in relation to specific kinds of such research. What, if any, research in the biological sciences is morally impermissible?

Research in the biological sciences undertaken for the purpose of weaponising biological agents so that they can be used to kill or cause illness in human populations is presumably morally impermissible, whether the research in question is undertaken by state actors, (non-state) terrorist groups, criminal organisations or malevolent individuals. So much is proclaimed in the Biological Weapons Convention (BWC), notwithstanding the fact that arguments have been used from time to time to justify the use of biological weapons in the context of a just war. It has been argued, for example, that some biological weapons are more "humane" than some conventional weapons. It has also been argued that biological weapons development during peacetime may play an important role in deterrence [79]. It is not within the scope of this report to discuss the moral complexities arising from the use of various forms of weaponry, albeit this is an important and somewhat neglected topic. However, we note that, in so far as biological weapons are a species of weapons of mass destruction (WMD), there is a *general* moral objection to their development and use, namely, that inevitably they target civilian populations and not merely combatants. As such, they violate the so-called *jus in bello* condition of just war theory; the condition that, among other things, gives expression to the moral principle of civilian immunity in war.

An analogue of the moral principle of civilian immunity in conventional wars between nation-states is the moral principle not to deliberately target civilians that is adhered to by some—but obviously not all—*non-state* actors engaged in armed struggles. For example, for most if not all of its history the African National Congress (ANC) in its armed struggle against the apartheid government in South Africa adhered to this moral principle; military and police personnel were regarded by the ANC as legitimate targets but not ordinary civilians. On the other hand, terrorist

S. Miller, M.J. Selgelid, *Ethical and Philosophical Consideration of the Dual-Use Dilemma in the Biological Sciences,* groups such as al Qaeda obviously violate this moral principle, as would any terrorist group using biological weapons as WMDs.[1] Naturally, terrorist groups might use "new generation" biological weapons that are able to target particular individuals, e.g., a biological weapon of assassination. However, use of such a biological weapon would not constitute use of a WMD.

In addition to the general concern that biological weapons may serve as WMDs is the concern that their effects are, generally-speaking, hard to predict and control. The fact that biological weapons are relatively inexpensive and easy to produce (in comparison with other WMDs) also means that the potential for an arms race in the context of biological weapons is especially worrisome. These features constituted central rationales behind the BWC.

At any rate, our assumption in this report is that research in the biological sciences undertaken for the ultimate purpose of stockpiling or using weaponised biological agents is in fact morally impermissible. The moral problem that now arises concerns research in the biological sciences that is not undertaken *by the original researchers* for the ultimate purpose of stockpiling or using weaponised biological agents, but might be used by secondary researchers (or other users) for these impermissible purposes, i.e., the moral problem presented by so-called dual-use dilemmas.

As already noted, a particularly morally problematic species of the dual-use dilemma arises in the case of research undertaken to enable the assessment of the potential threat posed by the biological weapons (BW) of other nation-states (including nation-states who might seek to use BW as instruments of terror against civilian populations) or the biological agent focused projects of non-state terrorist groups. Such 'threat assessment' research involves experimenting with the offensive applications of pathogens so as to determine appropriate counter-measures. In order to develop defences against a putative BW agent, it is necessary to understand:

- The underlying mechanisms for pathogenicity, including infectivity and virulence
- The way in which a micro-organism evades the human immune system or acquires resistance to antibiotics and
- The ways in which the agent may be dispersed, and its infectivity by each route

However, an understanding of these factors is also exactly what would be required for the development of BW [72]. An analogous point can be made in relation to non-state terrorist groups engaged in, for example, developing improvised equipment that could be used to grow a biological agent.

In relation to the dual-use dilemma in the biological sciences, the approach of the US National Research Council (NRC) in its influential 2004 report, *Biotechnology Research in an Age of Terrorism*, is to map the range of these dual-use dilemmas by identifying and taxonomising a set of salient "experiments of concern". We accept this approach in the context of our attempt to isolate the morally permissible from the morally impermissible in relation to dual-use research

[1] See Miller [86].

in the biological sciences. Our first task, then, is to map the terrain of such dual-use dilemmas; hence, our recourse to experiments of concern.

According to the NRC report "experiments of concern" are those that would:

1. Demonstrate how to render a vaccine ineffective
2. Confer resistance to therapeutically useful antibiotics or antiviral agents
3. Enhance the virulence of a pathogen or render a non-pathogen virulent
4. Increase the transmissibility of a pathogen
5. Alter the host range of a pathogen
6. Enable the evasion of diagnosis and/or detection by established methods or
7. Enable the weaponization of a biological agent or toxin [102].

Other possible categories are:

8. Genetic sequencing of pathogens
9. Synthesis of pathogenic micro-organisms
10. Any experiment with *variola* virus (smallpox)
11. Attempts to recover/revive past pathogens.

Experiments of Concern

Demonstrate How to Render a Vaccine Ineffective

The Dual-use Dilemma: It may be important to know, for public health reasons, whether and/or how a particular vaccine can become ineffective so that the vaccine can be improved or alternative protective measures can be investigated and adopted. The deployment of vaccine-resistant biological agents against a target population, however, would circumvent an important medical defence.

Examples: *Vaccine-resistant mousepox*. In 2001 the *Journal of Virology* published an article describing the accidental discovery of a group of Australian scientists who were attempting to produce an infectious contraceptive for mice, which periodically breed out of control in parts of Australia. The scientists spliced a single foreign gene into a mild mousepox virus in the hope of creating a genetically engineered sterility treatment. The gene—interleukin-4 (IL-4)—helps regulate immune system reactions. The effect, however, was to create a strain of mousepox so virulent that it killed both mice with natural resistance to mousepox and mice that had been vaccinated against mousepox [26]. A disturbing implication of this result is that adding an IL-4 gene might similarly increase the virulence of smallpox (or some other poxvirus that infects humans) and potentially allow the virus to overcome vaccination (which is our only defense against smallpox). The genetic engineering technique used in this study is relatively straightforward and described in standard microbiology textbooks. No extraordinary equipment or facilities are required. To employ the technique on smallpox, however, one would need access to the smallpox virus (which, officially anyway, only exists in two secure facilities worldwide).

Project Jefferson. In September 2001 the *New York Times* revealed the existence of a classified US biodefence project (Project Jefferson) which, in early 2001, involved the production of a vaccine-resistant strain of anthrax bacteria [152]. The purpose was to reproduce results of Russian research published by *Vaccine* in 1997. The researchers inserted genes from *B cereus* into *B anthracis* and showed that the engineered bacteria were highly lethal against hamsters, even when they had been inoculated with Russia's standard anthrax vaccine [44, 142]. The US officials involved in Project Jefferson were reportedly mindful of the BWC and the need for protective intent. Accordingly, the project aimed to produce only small quantities—1 g or less—of the modified anthrax [85, p. 309]. Though the Soviets allegedly had the capacity to produce 4,500 metric tons of anthrax yearly [85, p. 254], strictly speaking even one gram of anthrax is a large quantity, capable of infecting thousands of people if a suitable dried spore preparation is made.

When Project Jefferson produced a vaccine-resistant, genetically modified biological agent, it was only verifying something that had already turned up in the scientific literature. It is a different matter to produce modified pathogens that no one, potential adversary or otherwise, has ever created.

Cowpox. Some novel pathogens only exist as disease threats because scientists created them. In 2003, for example, a team of US scientists at St Louis University led by Mark Buller, supported by a National Institute of Allergy and Infectious Diseases (NIAID) biodefence grant, repeated a previously-published Australian experiment on mousepox (described above) [26] with the intention of developing a pharmaceutical countermeasure. In the experiment, mice infected with genetically-modified mousepox virus recovered when treated with a combination of antiviral drugs. As mousepox is closely related to the *variola major* (smallpox) virus, the result led Buller's team to hope that a treatment against genetically engineered smallpox could be developed [121]. Later, however, the scientists went further by applying the mousepox enhancement technique to the cowpox virus which, unlike mousepox, infects humans. The rationale was reportedly "[t]o better understand how easy or difficult it would be to apply the same kind of genetic engineering to the human smallpox virus and make it more lethal" [140]. Although such work has been justified as "necessary to explore what bioterrorists might do", other scientists have questioned the utility and wisdom of enhancing viruses [116, 157].

Confer Resistance to Therapeutically Useful Antibiotics or Antiviral Agents

The Dual-use Dilemma: The production of a pathogen which is resistant to existing treatments can be for genuinely peaceful purposes. A scientist might, for example, set out deliberately to generate antibiotic-resistant bacteria to determine whether or not, or how, a bacterial strain *can* become resistant. Such information would be relevant to recommendations about how best to administer the antibiotic; and resistant bacteria could be used to test alternative and/or new antibiotics.

Researchers may additionally study, and thus select for, resistant microbes in order to demonstrate the activity of an antimicrobial, to confirm the mechanism of an antimicrobial, and to discover the functions of microbial proteins. For these and other reasons, scientists routinely conduct experiments to assess the ability of pathogenic micro-organisms to acquire resistance (and produce resistant pathogens in the process): "[i]n vitro and in vivo selection for drug resistance has become part of the standard of all drug characterization and development" [51]. Drug resistant microbes able to defeat the defences that are erected by the human immune system and supplemented by existing medical technology, however, could make attractive biological weapons.

Examples: *Drug-resistant mousepox.* Given the perceived need, especially in light of the bioterrorist threat, for treatment against poxviruses such as smallpox, camelpox, and monkeypox, Australian scientists tested the efficacy of the antiviral cidofovir (which had previously proven effective against poxviruses in animal models) in the protection of varieties of mice against varieties of the mousepox virus [52]. While the drug was generally effective in protecting mice against mousepox, it was found that the more virulent strain of mousepox created by genetic engineering insertion of the IL-4 gene into the mousepox (*ectromelia*) virus (as described above) was resistant to the anti-viral effects of the drug. This might indicate that other drugs should be investigated and/or developed for protection against genetically engineered poxviruses. But it also reveals (to would-be bioterrorists) that insertion of the IL-4 gene into poxviruses may produce super-strains of disease that are drug-resistant (as well as possibly vaccine-resistant, as described above).

Drug-resistant plague and anthrax. The Soviet biological weapons program allegedly led to development of a strain of anthrax resistant to five antibiotics and a strain of plague resistant to all antibiotics. Insofar as this was part of an offensive biological weapons program—and because the purpose of the original researchers was malign—these are clear examples of impermissible research. A hypothetical example of dual-use research, however, would be the attempt to reproduce such pathogens for bio-protective purposes. If it is confirmed that creation of such pathogens is possible, and if the characteristics of such pathogens are determined, then we will know more about the kinds of threats (both natural and unnatural) we may need to prepare to protect ourselves against. The results of such research, however, could potentially be used by bioterrorists or others with harmful intentions.

Enhance the Virulence of a Pathogen or Render a Non-pathogen Virulent

The Dual-use Dilemma: For public health reasons, it may be important to know whether and/or how the virulence of a pathogen that exists in nature can increase. An 'enhanced' pathogen deployed in a biological attack, on the other hand, would inflict more human damage than normal.

Examples: In 2002 the *Proceedings of the National Academy of Sciences* (*PNAS*) published the results of an experiment involving the engineering, from published DNA sequences, of a protein—known as SPICE—produced by the smallpox virus. The study revealed the ways in which, and the extent to which, this protein defeats the human immune system. This could be important knowledge in the event that smallpox re-emerges because "[d]isabling SPICE may be therapeutically useful" [53]. But the results potentially provide information on how to increase the virulence of the closely related *vaccinia* virus (which is used in the smallpox vaccine).

In 2003 *PNAS* published the findings of scientists who had accidentally created a more virulent form of tuberculosis (TB) when trying to alter the genetic structure of *Mycobacterium tuberculosis* bacteria [60].

Increase the Transmissibility of a Pathogen

The Dual-use Dilemma: For treatment and public health planning purposes, it may be important to know whether a naturally-occurring infectious disease threat could be worsened by the evolution of a pathogen into a more transmissible form. Attempts to enhance transmissibility might thus yield valuable knowledge. A pathogen might be more useful as a biological weapon, however, if it is more easily transmitted through a population.

Examples: *H5N1 influenza.* The World Health Organization (WHO) has sponsored research to find out whether H5N1 avian influenza could trigger a human pandemic. The hope is that, by "reassorting" (mixing) H5N1 with human influenza viruses in the laboratory, scientists may determine how dangerous the hybrid virus would be and the likelihood of it causing a pandemic. Such experiments could help determine whether there is some natural barrier to the reassortment of H5N1 or whether the world has simply been lucky [19].

Switching the virus between different hosts is also believed to lead to increased transmissibility, especially if one of those hosts is the pig. A number of influenza laboratories are considering mixing birds with pigs to determine if the virus will mutate to increase its transmissibility, and potentially be of greater threat to humans.

Smallpox-Ebola chimera. A project of the Soviet bioweapons project allegedly aimed to produce a hybrid "chimera" of smallpox and Ebola. The purpose was to create a pathogen highly contagious like the former and highly virulent like the latter. It can thus be characterised as a project which aimed to increase the transmissibility of Ebola. (Such research would also fall under the third category of concern insofar as its purpose was to increase the virulence of smallpox.) The Soviet project should not be considered dual-use; insofar as this was conducted as part of an offensive biological weapons program this is clearly impermissible research. A hypothetical example of dual-use research, however, would be the attempt to construct such a chimera with the protective aim of discovering whether or not the Soviets or others may have succeeded in such a project. If it is discovered that creation of such a chimera is possible, and if the characteristics of such a pathogen are determined, then we will know more about the kinds of things we need to prepare to protect ourselves against. The knowledge about how to create such a virus, on the other hand, could be used to cause extreme devastation by those with malevolent intentions.

Alter the Host Range of a Pathogen

The Dual-use Dilemma: In an era of renewed concern about emerging infectious diseases, it may be important for human health to know whether a non-zoonotic disease can become, or is close to becoming, a zoonotic agent. Medically important research might thus result in an animal disease that could sicken humans. The use of such an agent in a biological attack could be devastating because, as was the case with the severe acute respiratory syndrome (SARS) coronavirus, people have no immunity and have not been selected for resistance to the disease.

Examples: An important area of research into infectious diseases examines why some pathogens infect multiple hosts whereas others are highly host-specific. By what processes, for example, did HIV become able to infect humans as well as chimpanzees? Understanding what determines host-specificity or host-limitation is important for research into emerging diseases of humans, animals and plants. An example of high risk research might be experiments to determine whether a species-specific virus such as camelpox could be adapted to readily transmit in humans by insertion of *variola* genes.

Myxoma virus was used for the biological control of rabbits and does not replicate in humans. However, genes are now being identified which when engineered into the *myxoma* virus could overcome host specificity and allow the virus to infect humans [27].

Enable the Evasion of Diagnosis and/or Detection by Established Methods

The Dual-use Dilemma: In relation to ongoing infectious disease threats to human health, it may be important to know whether a pathogen has the potential to mutate naturally into an undetectable form so that new diagnostic/detection techniques may be devised. Pathogens engineered to evade diagnosis and/or detection, however, would be well-suited for a covert biological attack; and the delay in diagnosis and subsequent treatment would increase the resulting human damage.

Examples: Microencapsulation of pathogen particles would be one way of avoiding antibody-based detection, although this technique has no analogue in nature. As such, microencapsulation would only be carried out for an offensive BW purpose (such as delivery of a pathogen to the lower intestine) or to investigate the requirements for protection against such a threat.

Altering gene sequences may be a way of testing the robustness of molecular detection systems. It may be useful to understand the circumstances under which natural mutations would be likely to render a diagnostic system ineffective.

Enable the Weaponization of a Biological Agent or Toxin

Experiments of this kind test the bounds of permissibility most severely. Weaponized agents do not exist in nature, and so (absent the threat of biological weapons attack) there is no ongoing public health imperative for protective mechanisms as there is against a naturally occurring infectious disease threat.

The Dual-use Dilemma: Understanding weaponization processes may facilitate the development of protections against a potential BW perpetrator (including a nation-state contemplating a terrorist attack on civilians). Our focus here will be on the weaponisation of biological agents by nation-states, as opposed to the processes for delivery of biological agents that might be used by non-state actors contemplating a terrorist attack. (We do not thereby mean to imply that the threat assessment in relation to the latter is not important; clearly it is of enormous importance.) Weaponization for "threat assessment" purposes is likely to be interpreted by outsiders as simply the production of BW, thus endangering the norm against their production, driving a biological arms race, and making biological attacks more likely.

Examples: *Project Clear Vision*. In September 2001, the *New York Times* revealed the existence of a classified US biodefence project (Project Clear Vision) which, from 1997 to 2000, involved building and testing a Soviet-model bomblet for dispersing bacteria [152]. This involved tests of bacteria bomblets, built according to a Soviet design, and conducted by Battelle, a military contractor in Columbus, Ohio. The bomblets were reportedly filled with simulant pathogens and tested for their dissemination characteristics and performance under different atmospheric conditions. Experiments in a wind tunnel revealed how the bomblets, after being released from a warhead, would fall on targets [85, p. 295]. Before the testing took place, some US government legal experts had argued the experiments were not a breach of the BWC provided they were not intended for offensive purposes. Other officials argued that a weapon was, by definition, meant to inflict harm and therefore crossed the boundary into offensive work: "A bomb was a bomb was a bomb" [85, p. 288].

Biological agent grenade. In February 2003 a patent was issued to the US Army for a "rifle-launched non-lethal cargo dispenser" that can be filled with "smoke, crowd control agents, *biological agents*, chemical agents, obscurants, marking agents, dyes and inks, chaffs and flakes" [emphasis added] [10]. In December 2005, after concerns were raised that the development and production of such a weapon would breach the BWC, the US Patent and Trade Office approved a change to the patent which removed the term 'chemical/biological agent' [122].

Aerosolization. Small-scale aerosolization technology may be useful for administering individual doses of inhaled vaccine or antiviral therapy (such as ribavirin) to humans, and larger-scale aerosolization could be used for mass-vaccination of animals—for example, in the poultry industry. It is hard to imagine large-scale aerosolization being therapeutically useful for humans, although such technology would certainly have enormous value for the purpose of delivering BW agents. Such technology might also be developed and tested for protective purposes. One of the principal aims of the NIAID Biodefense Research Agenda, for example, is to "ensure adequate numbers of BSL-3 [Biosafety Level Three] facilities with aerosol challenge capacity" [161, p. 8].

NBACC. The National Biodefense Analysis and Countermeasures Center (NBACC), due to be completed in 2008, is intended to provide the United States with high-containment laboratory space for biological threat characterization and bioforensic research. According to the US Department of Homeland Security, NBACC will form part of the National Interagency Biodefense Campus at Fort Detrick, Maryland. Its programs will investigate the infectious properties of biological agents, the effectiveness of countermeasures, decontamination procedures, and forensic analysis. Part of NBACC is the Biological Threat Characterization Center, which will conduct laboratory experiments aimed at investigating current and future biological threats. The Center will also assess vulnerabilities, conduct risk assessments, and determine potential impacts in order to guide the development of countermeasures such as detectors, vaccines, drugs, and decontamination technologies [135].

Many of the activities to be undertaken by NBACC could readily be interpreted by outsiders as the development of BW under the guise of threat assessment. In particular, weaponization projects and the construction of novel (not previously existing) pathogens arguably constitute impermissible research. In a February 2004 presentation, George Korch, Deputy Director of NBACC, revealed that one of its research units intended to pursue a range of topics including "aerosol dynamics", "novel packaging", "novel delivery of threat", "genetic engineering", and "red teaming". At one point in his presentation, Korch summarized the threat assessment task areas as: "Acquire, Grow, Modify, Store, Stabilize, Package, Disperse" [38, 115, 117]. Such language is identical to that which would describe the functions of an offensive BW program.

Indeed, a 1998 report from the Office of the US Under Secretary of Defense for Acquisition and Technology stated: "Stabilization and dispersion are proliferation concerns because these technologies increase the efficacy of biological agents" [163]. And in light of the planned NBACC activities as described by Korch, a 2005 US State Department report which assessed that "China maintains some elements of an offensive BW capability in violation of its BWC obligations" appeared to reflect an American double standard on BW when it warned that:

> From 1993 to the present, [Chinese] military scientists have published in open literature the results of studies of aerosol stability of bacteria, models of infectious virus aerosols, and detection of aerosolized viruses using polymerase chain reaction technology. Such advanced biotechnology techniques could be applicable to the development of offensive BW agents and weapons [136].

Genetic Sequencing of Pathogenic Micro-Organisms

The Dual-use Dilemma: Sequencing the genetic codes of entire pathogens or specific genes of pathogens could assist in understanding the nature of the pathogens and in the development of new vaccines or treatments for the diseases they cause. Sequencing also facilitates genetic diagnosis and detection. Gene sequence data could, on the other hand, be used to construct a pathogen for deployment against a target population with no natural immunity. Of particular concern are the facts that the smallpox genome has been published and that the published polio genome enabled artificial synthesis of a "live" polio virus as described below.

Examples: *Anthrax*. In a letter to *Nature*, Read et al. describe the sequencing of the Ames strain of *Bacillus anthracis* (anthrax). Reported benefits of the sequencing include identification of (1) virulence genes on plasmids, (2) "chromosomally encoded proteins that may contribute to pathogenicity", and (3) "numerous surface proteins that might be important targets for vaccines and drugs" [49]. They conclude that "the complete sequence of *B. anthracis* is a step towards a better understanding of anthrax pathogenesis." Though presumably true, the improved understanding of anthrax enabled by genetic sequencing could potentially be used by those who aim to increase the danger of anthrax—or to transfer its harmful characteristics to other microbes (including near genetic neighbours identified by the study) via genetic engineering—as well as those who aim to improve medical defenses against it.

Influenza. Research results published in 2005—on the complete genetic sequencing of the 1918 influenza A (H1N1) virus [63] and the "resurrection" of H1N1 using reverse genetic techniques [65]—revealed (and reproduced, in animals at least) the traits that made the virus so virulent. However, the decision to publish this information aroused concerns that would-be BW perpetrators could use it to reconstruct H1N1 for malign purposes. The danger of the virus is revealed by the fact that it killed between 20 and 100 million people in 1918–1919—more than have been killed by any single disease in such a short time period in human history. The newly created US National Science Advisory Board for Biosecurity (NSABB) was asked to consider the latter paper before publication and concluded that the scientific benefit of the future use of this information on the 1918 virus far outweighed the potential risk of misuse. Similar issues relate to publication of the H5N1 influenza—i.e., bird flu—genome.

Synthesis of Pathogenic Micro-Organisms

The Dual-use Dilemma: Synthesis of the genomes of viruses theoretically allows the introduction (and precise positioning) of mutations or novel sequences that can be used to study the function of particular genes or regulatory sequences. It would be more usual (though perhaps less precise) to do this using conventional molecular biology—e.g., "infectious clones"—rather than genome synthesis [47]. Synthesis technology would obviate the need for would-be BW perpetrators to source pathogens from natural reservoirs in other parts of the world or from other laboratories. It can facilitate reconstruction of extinct pathogens (as mentioned above and below, with regard to the "resurrection" of the 1918 flu) and it could theoretically enable construction of novel pathogens.

Example: *Polio*. In an experiment carried out partly with the intention to draw attention to BW threats, American scientists sponsored by the US Department of Defence spent three years synthesising a *poliomyelitis* (polio) virus "from scratch". Following the map of the polio virus RNA genome published on the internet, they bought and strung together corresponding DNA sequences, commercially available over the internet. This was used in a cell-free extract to create "live" virus that paralysed and killed mice [9]. One of the polio project scientists, Eckard Wimmer, said the experiment proved that eradicating a virus in the wild might not mean it is gone forever—conceivably, scientists may soon be able to apply the same technique to synthesise more complex viruses such as Ebola using blueprints available in scientific archives and from biological supplies purchased through the mail [36, 125, 137]. He said they "made the virus to send a warning that terrorists might be able to make biological weapons without obtaining a natural virus" [156]. Of particular concern is the possibility that the technique would enable artificial synthesis of smallpox. One reason the technical feasibility of the latter is doubtful, however, is the fact that the smallpox genome is so much larger—i.e., 200,000 base pairs in comparison with 7,500 for polio. Adding to such prospects, however, in December 2004 *Nature* described an unexpectedly sudden advance in synthesising longer strands of DNA: a research team synthesised a molecule 14,500 chemical units in length [46, 64]. While it took a number of years to initially synthesise polio, furthermore, Craig Venter has succeeded in synthesizing bacteriophages of comparable size (i.e., 6,000 base pairs) in a matter of weeks. This technology is advancing so rapidly that it is difficult to predict what will be possible in the future. Many now believe that synthesis of smallpox is possible.

Any Experiment with the Smallpox Virus

The smallpox virus (*variola major*) is a special case because it no longer exists in nature. As such, there is no public health imperative to defend against naturally occurring smallpox outbreaks. On 8 May 1980 the World Health Assembly (WHA) declared the successful eradication of smallpox worldwide, and by 1984 the World Health Organization (WHO) had authorized only the United States and the Soviet Union to each possess a single repository of the world's last samples of the virus.[2] In May 1990, US Health and Human Services Secretary Louis Sullivan proposed that the two countries should start working towards eliminating their collections of *variola* in order to ease fears that smallpox might be used as a biological weapon [92]. This sparked a debate over destruction versus retention of the virus that has continued to this day.

The Dual-use Dilemma: Understanding *variola* is important for developing medical defences in the event that a smallpox outbreak occurs as a result of a BW attack or the accidental leak of the virus from a laboratory. Because biosafety and biosecurity measures in laboratories are less than perfect, however, an increase in the number of personnel and facilities working with *variola* increases the likelihood of the virus escaping or being stolen and used in a biological attack.

Example: The WHA in 1999 established the Variola Advisory Committee (VAC). At its meeting in November 2005 the VAC noted, *inter alia*, further work on the primate model of human smallpox, which had been undertaken to facilitate the development of antiviral drugs and to meet associated licensing requirements. Recent experiments in primates, using different doses of virus, had induced disease with features similar to that of lesional smallpox and haemorrhagic smallpox in humans. These studies were considered useful in assessing the efficacy of antiviral drugs because they enabled greater understanding of specific organ and tissue sites of viral replication [164]. The relevance of this model is seriously questioned because very high levels of smallpox virus challenge were utilised, 10^9 for aerosol and 10^8 for intravenous challenges.

Attempts to Recover/Revive Past Pathogens

The Dual-use Dilemma: A number of attempts have been made to recover past pathogens from preserved frozen bodies. While the recovery of such pathogens may reveal important historical, evolutionary, and medical information, such patho-

---

[2] The samples are presently stored in secure freezers at the Centers for Disease Control and Prevention (CDC) in Atlanta and at the Centre for Research on Virology and Biotechnology in Koltsovo, Russia.

gens may be extremely dangerous to human populations if they are accidentally or intentionally released into the environment.

Examples: Journalist Gina Kolata describes numerous examples of successful and failed attempts by scientists to recover the 1918 flu from the lungs of frozen bodies in permafrost in Alaska and Norway [83]. Some of this research involved remarkably little oversight and precaution. One risk of this research is the possibility that the researchers themselves would be infected when exposed to the bodies in question—and the possibility that this could have sparked a global epidemic. More relevant to the dual-use dilemma is the fact that some of these researchers aimed to revive the frozen virus. While revival attempts have failed, recovery of and research on the 1918 flu virus from frozen bodies enabled the genetic sequencing and "resurrection" via reverse engineering discussed above. Both the revived virus, if attempts had been successful, and the information gained from the recovered frozen 1918 (HIN1) flu virus could be used for both beneficial and harmful purposes. Kolata also cites reports that Russian scientists have attempted to recover smallpox from frozen bodies. Attempts to recover/revive (particular strains of) other preserved pathogens are easily imaginable.

Chapter 3
The Ethics of Dual-Use Research

By definition, dual-use research is morally problematic. On the one hand, such research provides benefits (at least potentially); on the other hand, there is the risk of misuse by rogue states, terrorists groups, and the like.

Broadly speaking, the most obvious benefits of research in the biological sciences of the kind in question are: the protection of human life and physical health against diseases (including novel ones), the protection of existing, and (more controversially) the provision of novel, food sources; and the protection of human populations against biological weapons. By contrast, the potential burdens of such research are death and sickness caused by the use of biological agents as weapons in the hands of malevolent state actors, terrorist groups, criminal organisations, or individuals.

More fine-grained analyses of the benefits and burdens of such research would elaborate on the additional kinds of benefit/burden and recipients/bearers thereof, e.g., the economic wealth accrued by large pharmaceutical corporations and their shareholders, the economic costs of expensive, unsuccessful (or only marginally beneficial) research programs in the biological sciences and, more generally, the dis/utility and in/justice of specific allocations of resources to, and the distribution of benefits and burdens from, different research programs in the biological sciences, e.g., the evident disutility of the large 1946–1992 Soviet biological weapons program.

Fine-grained ethical analyses of dual-use research in the biological sciences would seek to *quantify* actual and potential benefits and burdens, and actual and potential recipients/bearers of these benefits and burdens. These analyses would also identify a range of salient policy options. Each option would embody a set of trade-offs between present and future benefits and burdens, and recipients and bearers thereof. The construction of these options and the process of selection between them would consist in large part in the application of various ethical principles, including human rights principles—e.g., right to life, freedom of inquiry, and free speech—and principles of utility and justice. Here we note that there is no simple inverse relationship between specific benefits and burdens such that, for example, any increase in security requires a reduction in scientific freedom. Rather, an increase in security might simply involve greater safety precautions and, therefore, a financial cost without any commensurate reduction in scientific freedom. At any rate, relevant

S. Miller, M.J. Selgelid, *Ethical and Philosophical Consideration of the Dual-Use Dilemma in the Biological Sciences*,

benefits and burdens need to be disaggregated and subjected to individual analysis in the context of any process of determining trade-offs and selecting options.

We cannot here provide such fine-grained ethical analysis, but will rather focus (somewhat simplistically) on a single ethical consideration, namely, human health (including human life) that gives rise to the dilemma; and do so without exploring questions of which human populations or how many individual humans have benefited/been burdened or are likely to benefit/be burdened, and so on. Viewed from this perspective the dual-use dilemma concerns human health (as a simple, unquantified human good), and the dilemma consists in the fact that research undertaken to promote human health might instead be used to destroy human health. As such, the dilemma gives rise to questions of security: what are reasonable and ethically justified forms and degrees of security in this context?

The security in question is a complex notion. It consists in part in the physical security of, for example, samples of biological agents against theft. Relatedly, security consists in part in the processes in place to ensure, for example, that the researchers themselves cannot, or will not, conduct research for malevolent purposes. As we will see in the next chapter, security in this sense also consists in part in restrictions that might be placed on the dissemination of research findings.

Thus far we have offered a somewhat static mode of analysis of the dual-use dilemma consisting of the quantification of harms and benefits, the identification of salient options, and the selection of an option on the basis of ethical principles. However, a more dynamic, indeed creative, mode of analysis is called for. In the first place, options are not static because well-intentioned scientists, malevolent actors and security personnel are *responsive* to the problems that they confront, including the problems posed by other actors. The response of scientists to a pathogen with enhanced virulence might be the development of a new vaccine. The response of security personnel to a new bioterrorist threat might be an enhanced regulatory system. Accordingly, the mode of analysis of the dual-use dilemma must be dynamic in character.

In the second place, ethical dilemmas are not necessarily—or even typically—to be resolved by careful calibration of the differential ethical weight that attaches to the options provided for in the dilemma. Rather the dilemma must, if possible, be resolved by *designing* a new third or fourth option, i.e., by bypassing the dilemma. Consider the question of whether to disseminate dual-use research findings or not disseminate them: academic freedom versus security. Perhaps the solution is to find a third option, such as to disseminate them in a manner that will not enable the experiments in question to be replicated (other than by those with adequate security clearance and to the extent necessary for purposes of verification). This mode of analysis is *creative*. It lets us have our cake and eat it; it squares the ethical circle. Let us refer to this kind of ethical analysis as *designing-in ethics* [67]. Naturally, this mode of analysis is not always applicable; but it is important to keep it in mind.

In the light of these considerations of health and security—and of this designing-in ethics mode of analysis—let us address the question of the moral permissibility of dual-use research, albeit in highly general terms. Here there appear to be three separable ethical questions. Firstly, the ethical question as to whether or not a putative biological agent to be researched ought in fact to be eliminated (or, if already eliminated, not retrieved). Here the *possibility* of research is removed; no possibility of research because no biological agent to be researched. We have in mind the case of smallpox and the arguments in favour or against the elimination of all samples of smallpox. (We note that in the context of the recent possibility of genome synthesis, it may not matter so much whether the organism actually exists; since it can be recreated from the sequence alone.) Secondly, the ethical question (or questions) arising from dual-use research in relation to a biological agent whose present and/or future existence is taken as a given; there is no intention to eliminate or not retrieve or not bring into existence the biological agent in question. For example, research to determine whether or not avian influenza could trigger a human pandemic might lead to the creation of dangerous new strains that could be used by terrorists. Such research might include work intentionally undertaken to create novel pathogens or synthesising existing ones, albeit work whose ultimate purpose was to develop, say, a vaccine against these pathogens. Thirdly, the ethical question of whether to undertake dual-use research for the purpose of protection against weaponised pathogens, e.g., research into the aerosolisation of pathogens.

Let us consider these ethical questions in order, beginning with the question of the elimination of pathogens. As already mentioned, the salient example here is smallpox. The "retentionists" focus on the possibility that smallpox might re-emerge in the future. They see potential sources for a return as including the corpses of smallpox victims preserved in Arctic permafrost, and samples of *variola* virus retained without World Health Organization (WHO) permission—inadvertently or deliberately—in laboratory freezers around the world. The former Soviet Union reportedly had enormous weapons stockpiles of the virus, for example, and their destruction has never been verified. There is also some concern that genetic mutation of the monkeypox virus, which causes symptoms clinically almost indistinguishable from smallpox, might make it more virulent and more transmissible, thus resembling the public health threat once posed by *variola*. Against these eventualities, it is thought prudent to study the smallpox virus in order to be better prepared. Another argument in favour of retention (and perhaps retrieval) is that basic research with live *variola* could yield new insights into the process of viral infection generally and the workings of the human immune system [74, 92, p. 171].

Among the 'destructionists' are many veterans of the smallpox eradication campaign, including D.A. Henderson and Frank Fenner. They argue that, because the DNA sequences of representative viral strains have been determined, there is no need to retain live *variola* in order to identify smallpox were it to reappear in the future. Moreover, they argue, live virus is not required to protect against a future outbreak because the smallpox vaccine—derived from the *vaccinia* virus—could be retained and stockpiled just in case. There is also the argument that eliminating the remaining laboratory stocks of *variola*, and universally criminalising mere possession of the virus, would strengthen the moral case against using smallpox as a biological weapon [25, 92, pp. 170–171].

The smallpox debate is essentially about balancing the desirability of retaining the virus for prospective research purposes against the uncertain risks associated with not destroying it. In an attempt to achieve this balance, the World Health Assembly (WHA) in 1999 established the Variola Advisory Committee (VAC). At its meeting in November 2005, the VAC recommended further experimentation on the live smallpox virus, including genetic modification. This proposal met with a cool reception when the WHA convened in May the following year. The WHO Director General, Lee Jong-Wook, rejected the recommendation to allow insertion of smallpox genes into related, less virulent poxviruses such as those that cause monkeypox and cowpox [139]. And the WHA urged the VAC to take extra steps to ensure laboratories had strong safety and security measures in place before smallpox research was carried out [141]. One good reason for such caution is that the deliberate release of modified smallpox virus or a poxvirus expressing smallpox genes could trigger a public health catastrophe.

Let us now consider the second ethical issue; it concerns dual-use research on presently existing or novel pathogens (where their present and/or future existence is taken as given). There are a number of types of experiment from our list of experiments of concern that are relevant to this question. However, the general problem here is the unintended (by the original researcher) untoward consequences of otherwise benign research. These consequences are threefold. First, there is an unintended dangerous biological research outcome, e.g., a pathogen with enhanced virulence or transmissibility or at least the knowledge of how to create such a pathogen. (As we have seen, there are some *intended* dangerous research outcomes, e.g., intentional creation of a vaccine resistant strains of a disease, undertaken for, say, prophylactic purposes, e.g., to test the adequacy of a vaccine, which have no untoward consequences.) Whether or not such an unintended and untoward outcome is possible or likely is a scientific question, best answered by biological scientists. Secondly, there is an outcome not intended by the original scientist but, nevertheless, intended by some malevolent state actor, non-state terrorist group, criminal organisation or individual, e.g., the weaponisation (and use as a weapon) of the pathogen that has been unintentionally created. Whether or not this outcome is possible or likely—given, say, a pathogen has already been (unintentionally) created—is a security question, best answered by security experts (with input from relevant non-security specialists such as engineers). Thirdly, there is the ultimate outcome intended by the malevolent individual or organisation, namely, the public health outcome of the biological attack. What the public health effects of a given biological attack are likely to be, e.g., in terms of morbidity and mortality, are public health questions, best answered by public health experts or teams thereof (including biological scientists, medical personnel and weapons experts, but also those knowledgeable about public health resources and infrastructure).

The danger attendant upon a given dual-use research program can be crudely quantified by determining the probability, be it low, medium or high, of a given untoward outcome, and multiplying this probability by the (quantified) disvalue (or disutility) of that outcome, e.g., in terms of the numerical loss of human life. A more fine-gained ethical analysis would explore the variety of decision making/

risk-taking strategies—including the precautionary principle[3]—that might be considered appropriate in this context. Presumably, dual-use research that has a high probability of resulting in substantial loss of human life ought not to be undertaken. On the other hand, the danger attendant upon dual-use research is not the only moral consideration in play. Another important moral consideration relates to the (intended) benefits of the research. Clearly trade-offs need to be made between the (intended) benefits of the research and its (unintended by researchers) potential untoward outcomes. Moreover, the process of arriving at suitable trade-offs is in large part a process of moral reasoning, including the weighing of one moral consideration against another. However, as noted above, it is important to bear in mind the possibility of creative solutions that bypass the dilemma; perhaps we do not need to make the trade-offs we initially think we need to make.

The general point to be made here is that in the context of the yet to be decided grey area of dual-use research marked off by the experiments of concern, there is a complex mix of scientific, security, public health and ethical considerations in play. Moreover, the process of moral reasoning involved will require trade-offs between ethical considerations and, hopefully, it will involve the provision of creative solutions that bypass the dilemma. The result will presumably be that some putative experiments of concern will be relegated to the impermissible category and others to the permissible category, albeit in the latter case under stringent conditions of safety and security.

Consider an experiment of concern involving enhancing the virulence of a pathogen. Susan Wright has argued, "[i]f there is no evidence of a threat posed by, say, a genetically engineered strain of cowpox that attacks the immune system, then there is no reasonable justification for developing such an organism. Arguably, to do so crosses the line between defence and offence" [140]. No doubt, pace Wright, there needs to be some evidence of a threat. But this raises a number of questions: What counts as evidence? How immediate is the threat? Does the development of the more virulent pathogen constitute a greater threat than the original threat that it is supposed to protect against? Surely when a microbial threat exists only in a scientist's imagination, an experiment to create such a microbe is both unnecessary and overly risky.

There are two additional points that should be stressed here. One pertains to the process of moral reasoning. We have been speaking in broadly utilitarian and consequentialist terms, e.g., using notions of future benefits/burdens, quantified loss of life, disvalue, and disutility. However, some would argue that this mode of reasoning is flawed (and not only by virtue of its inherently static character—see above). For example, consequentialist reasoning is arguably one-dimensional and fails to give sufficient weight to the intrinsic moral properties of current actions, e.g., perhaps human rights of current persons override future utility. Again, there are a

[3] Roughly, the principle to the effect that if research might cause great harm then, in the absence of a scientific consensus that the harm would not ensue, then the research should not be pursued. See, for example, Weckert and Moor [68].

range of moral considerations that are absent from our discussion thus far, e.g., the human right to free inquiry and intellectual property rights (both individual and collective). We do not have the space here to unravel all these moral complexities, or to develop and defend our own favoured account of moral reasoning as it might apply to the dual-use dilemmas in question. Accordingly, we have simply sought to gesture at some of the moral considerations in play and at the general kind of process of moral reasoning that should take place.

Another point pertains to uncertainty. Proceeding in the manner of a risk assessor assumes that the probabilities of specific outcomes can realistically be determined; risk assessment is, presumably, more than mere guesswork. But the reliability of probability judgments in relation to outcomes from dual-use research in the biological sciences is, to say the least, open to question. Arguably, the possibility of the development of a vaccine-resistant strain of smallpox based on research undertaken on mousepox to develop a contraceptive for mice could not have been realistically predicted. This is, of course, not to say that attempts should not be made to foresee untoward outcomes; it is merely to caution against overconfidence in the results of such attempts. Moreover, because the actors involved in dual-use dilemmas are, as noted above, responsive to problems and to one another's actions, probability judgments need to take this into account. One way to do so is to analyse, for example, a security risk from bioterrorists in part in terms of a complex set of variables including the ability, opportunity and motivation of the bioterrorists, the likely intelligence possessed by, and the likely assessment made by, the terrorists, the capacity to respond to specific forms of bioterrorist attack, the likely movements of innocent third parties at risk from specific security responses, the relevant moral principles, the rights and duties of the various actors involved, and so on.

However, what might be crucial here is the capacity to generate a creative response to the security problem thus analysed. Perhaps a focus on reducing the opportunities available to bioterrorists by establishing a licensing system for laboratories using dual-use technologies, licensing of DNA synthesisers and checks in relation to those who buy or use them are cases in point. Moreover, the developments in synthetic genomics may call for an adjustment in relation to existing security arrangements. For example, novel pathogens may escape classification under the US Select Agent regulations. Indeed, a new type of classificatory system might need to be introduced; one that makes use of functional or causal definitions of agents, instead of more traditional taxonomic systems.

Our third ethical question arising from dual-use research pertains to weaponisation. As already noted, a particularly morally problematic species of the dual-use dilemma arises in the case of research and development (R&D) projects on biological weapons (BW). In order to develop defences against a putative BW agent, it is necessary to understand the underlying mechanisms for pathogenicity and the ways in which the biological agent may be dispersed. However, an understanding of these factors is also exactly what would be required for the development of BW.

As stated above, our assumption in this report is that the weaponisation of pathogens for offensive military purposes is morally impermissible. Moreover, in so far as military defence is understood to include using biological weapons against an attacker (whether the attack in question is a biological attack or not) the weaponisation and/or stockpiling of pathogens for defensive military purposes—in this sense of defence—is also morally impermissible. On the other hand, research that is defensive in the sense that it serves the purpose simply of enabling combatants and civilians to *protect* themselves against a biological attack by, for example, developing early warning indicators of the presence of aerosolised novel pathogens is *prima facie* morally permissible. Let us refer to such research as research undertaken for the purpose of *protection* (as opposed to military *defence*). The problem is that such research for protective purposes might itself involve, for example, the creation of a virulent and highly transmissible novel pathogen and the weaponisation of it. As such, the weaponisation of pathogens for protective purposes gives rise to a dual-use dilemma of a very acute kind. The issue resolves itself into whether or not in practice the weaponisation of pathogens for protective purposes can be distinguished from the weaponisation of pathogens for offensive purposes (including defensive purposes in the above adumbrated sense of that term).

Presumably, if these two conceptually distinct activities are to be distinguished in practice then this is because there are *verifiable* differences in respect of: (i) intention or purpose; and/or (ii) physical properties of the weaponised pathogen. Intentions are, of course, extremely difficult to verify in practice. If the pathogen in question is one that some identified malevolent state or terrorist group is known to have weaponised, then the claim that a given body of research is protective may be plausible. But so long as there is reason to believe that some terrorist group might aim to weaponise the pathogen, or that the pathogen could be weaponised by someone, the claim that a given body of research is protective may be difficult to discredit.

A focus on physical properties is perhaps more promising. If the weaponised pathogen is possessed in large quantities, i.e., quantities appropriate for a military offensive but unnecessary for research serving purely protective purposes, then it is a case of weaponisation for offensive purposes. Unfortunately, recourse to quantities may not always resolve this question in the case of biological agents. Also one might expect there to be some differences in the *results* of research involving weapons constructed in accordance with the design-purpose only of testing protections against an attack using that weapon from the *results* of research undertaken with the design-purpose of making a successful attack using that weapon, i.e., an attack against which the enemy is not protected. For example, in the case of the former the research result might be a protective vaccine, whereas in the case of the latter the research result might be a weaponised pathogen that is resistant to any vaccine.

Moreover, the *results* of such dual-use research on the weaponisation of pathogens undertaken only for protective purposes, e.g., the vaccine mentioned above, might be more likely to be disseminated; after all, it is not only one's own civilians and combatants that ought to be protected from biological attack. Or at least one would expect, other things being equal, there to be less need for secrecy in relation to such relatively benign research and a willingness on the part of those engaged in it to be subject to verification checks.[4] US censorship of research on the medical use of penicillin during World War II, however, aimed to deprive the Germans of the benefits of what was then a new "wonder drug" [79]. History thus shows that the willingness to share medically beneficial research results cannot always be counted on.

[4] Naturally, other things might not be equal. We might be concerned that if a malevolent state knew what our defensive capabilities were they would be more likely to develop new ways to overcome them.

Chapter 4
Dissemination of Dual-Use Research Results

A primary area of contention in the context of dual-use science surrounds questions about whether or not, or the extent to which, restriction on dissemination of information gleaned from dangerous discoveries is warranted. This is an important issue in the life sciences in particular, because the tradition of information sharing in the life sciences has historically been almost completely open, especially in comparison with nuclear science, where discoveries with implications for weapons-making are automatically "born classified". Ironically, because biological weapons are so much easier to make than nuclear weapons—with regard to the expertise, expense, equipment, and materials required—one might think that secrecy and restriction on information dissemination would have been more important in the former context than the latter. However, it might be that because biological weapons are easier to make secrecy and restriction on information dissemination is less possible than in the case of nuclear weapons. Also secrecy might interfere with the development of counter measures.

A reason there has not been more control over dissemination of information in the life sciences is presumably that the dual-use phenomena is the clearest and strongest in this area of science. In biology it is more often *the very same discovery* that has both a direct beneficial purpose (with regard to medicine, for example) and a potentially harmful purpose (with regard to weapons-making). Nuclear discoveries related to weapons-making do not so often have such a direct and obvious link to human benefit (aside from the potential of nuclear weapons to be used for deterrent and/or defensive purposes).

While the protection of security and public health may arguably provide grounds for limiting dissemination of information related to dual-use discoveries, at least in certain instances, the issue of censorship should not be taken lightly. Governmental control over dissemination of information poses threats to widely cherished goods such as academic freedom (of inquiry), scientific autonomy, and freedom of speech itself. It is commonly held that these things are not only good in themselves but that they are also essential to the progress of science. Governmental control over science has an unfortunate history illustrated by examples ranging from Galileo in renaissance Europe to Lyshenko in the former Soviet Union. Other examples include the Biopreparat program in the USSR and the history of research on biological weapons in general.

S. Miller, M.J. Selgelid, *Ethical and Philosophical Consideration of the Dual-Use Dilemma in the Biological Sciences*,

Scientific openness and the free sharing of information are important to the methodology of the scientific enterprise as a whole. In response to claims that the mousepox and polio studies and others like them should not have been published, or to claims that the materials and methods sections of such articles should have been altered or omitted, for example, defenders of publication point out the importance of recognizing the extent to which a discovery in one area of science may have profound implications for progress in other areas. Because limiting description of materials and methods would interfere with processes considered essential to science—i.e., replication and verification—it was objected that such a practice would be at odds with the way that science actually works.

For these and other reasons—including the commonly held belief among scientists that knowledge is good in-and-of itself [81]—many in the scientific community strongly believe that things like secrecy and/or governmental control over science is contrary to what science is about. According to Robert Oppenheimer, who provided the scientific leadership of the Manhattan project which produced the first atomic weapons in the United States, "Secrecy strikes at the very root of what science is, and what science is for" [90]. From this perspective, even self-censorship is problematic from a scientific standpoint.

Academic Freedom

Oppenheimer's argument invokes academic freedom and, by implication, the human right to freedom of intellectual inquiry. It is worth spelling out some of the ethical issues here. In what follows we provide a brief analysis of academic freedom. Naturally, we do so in the context of the assumptions we have already made in relation to permissible and impermissible research. If a particular university-based research program or experiment of concern is morally *impermissible* given the safety, security and health concerns outlined above, then the moral principle of academic freedom has been rightly overridden; academic freedom is an important moral value, but it is not an *absolute* value.

The argument for the principle of academic freedom begins with the premise that freedom of intellectual inquiry is a fundamental human right [86, pp. 197–202]. Thus conceived, freedom of intellectual inquiry is not an individual right of the ordinary kind. Although it is a right which attaches to individuals, as opposed to groups, it is not a right which an individual could exercise by him/herself. Communication, discussion and inter-subjective methods of testing are social, or at least interpersonal, activities. However, it is important to stress that they are not activities which are relativised to social or ethnic or political groups; in principle, intellectual interaction can and ought to be allowed to take place between individuals irrespective of whether they belong to the same social, ethnic or political group. In short, freedom of intellectual inquiry, or at least its constituent elements, is a fundamental *human right*. Note that being a fundamental human right it can, at least in principle, sometimes override collective interests and goals including organisational, and even national, economic interests and goals. This "trumping" property of human rights is a constitutive element of liberal democracy; a form of polity whose legitimacy is based in part on its capacity and willingness to protect human rights including, at times, against infringements emanating from the government of the day.

If freedom of intellectual inquiry is a human right then like other human rights, such as the right to life and to freedom of the person, it is a right which academics as humans possess along with all other citizens. But how does this bear upon the specific institutional purpose of the university to acquire, transmit and disseminate knowledge?

Before we can answer this question we need to get clearer on the relationship between the human right to freely engage in intellectual inquiry on the one hand, and knowledge or truth on the other. Freedom of intellectual inquiry and knowledge are not simply related as means to end, but also conceptually. To freely inquire is to seek the truth by reasoning. Truth is not an external contingently connected end which some inquiries might be directed towards if the inquirer happened to have an interest in truth, rather than, say, an interest in falsity. Rather truth is internally connected to intellectual inquiry. An inquiry which did not aim at the truth would not be an intellectual inquiry, or at least would be defective qua intellectual inquiry. Moreover, here aiming at truth is aiming at truth as an end in itself. (This is not inconsistent with also aiming at truth as a means to some other end.) Further, to engage in free intellectual inquiry in our extended sense involving communication with, and testing by, others, is to freely seek the truth by reasoning with others. Intellectual inquiry in this sense is not exclusively the activity of a solitary individual.

Given that freedom of intellectual inquiry is a human right, and given the above described relationship between intellectual inquiry and truth (or knowledge) we can now present the argument in relation to freedom of intellectual inquiry. This argument in effect seeks to recast the notion of freedom of intellectual inquiry in order to bring out the potential significance, for conceptions of the university, of the claim that freedom of intellectual inquiry is a human right.

(1) Freedom of intellectual inquiry is a human right.
(2) Freedom of intellectual inquiry is (principally) freedom to seek the truth by reasoning with others.
(3) Freedom to seek the truth by reasoning with others is a fundamental human right.

Let us grant the existence of a human right to freely pursue the truth by reasoning with others. What are the implications of this right for universities and for academics' freedom of inquiry? Given such a right of intellectual inquiry, it is plausible to conclude that the university is simply the institutional embodiment of that moral right. In short, the university is the institutional embodiment of the right to freely seek the truth by reasoning with others.

The following claims now seem warranted:

First, universities have been established as centres wherein independence of intellectual inquiry is maintained. This flows from the proposition that the university is an institutional embodiment of the moral *right* of the inquirers to freely undertake their intellectual inquiries. Universities are not, for example, research centres set up to pursue quite specific intellectual inquiries determined by their external funders. Nor should particular inquiries undertaken by academics at universities be terminated on the grounds that some external powerful group, say government, might not find the truths discovered in the course of these inquiries politically palatable.

Second, universities have a duty to disseminate scholarship and research to the community. Intellectual inquiry is not only a human right, it is an activity which produces external benefits. For example, knowledge is a means to other goods, including economic well-being. Accordingly, and notwithstanding the rights of academics to freely inquire, it is reasonable that, qua community supported institutions, universities take on an obligation to ensure that their intellectual activities have a flow through effect to the wider community in terms of such external benefits. Thus dissemination of research (usually) has obvious benefits to the community, including health and economic benefits.

On the view of the university under consideration, interference in the process of the free pursuit of knowledge in universities strikes at one of the fundamental purposes for which universities have been established. Such interference could not be justified, for example, on the grounds that whereas free inquiry might be necessary for the acquisition of knowledge in many instances, in some particular instance free inquiry was not leading to knowledge, and therefore in this case free inquiry could be interfered with without striking at the basic purposes of the university as an institution.

Moreover, the university, in so far as it pursues this purpose, can so pursue it, even if so doing is inconsistent with the collective goals and interests of the community or government. In this respect the right of intellectuals to pursue the truth is akin to the right of the judiciary to pursue justice even in the face of conflicting collective goals and interests, including the national interest. The Mabo decision of the High Court of Australia to uphold certain land rights of Australian aboriginals might prove not to be in Australia's national economic interest. But in so far as judges were entitled as a matter of law and justice to recognise native title they were entitled to make the decision that they made. Similarly, Australian academics researching political or ethical issues in say, China or Indonesia, have a right to publish that research notwithstanding the damage it might do to present diplomatic relations and economic prospects.

Notwithstanding the importance of the human right of intellectual inquiry and its centrality to the institution of the university, freedom of intellectual inquiry in general, and of scientific inquiry in particular is not—as noted above—an absolute right. Specifically, it can be overridden if its exercise comes into conflict with other human rights, notably the right to life. Accordingly, if a contingency arose, such as war or a pandemic or a potential terrorist attack, then the duty of a scientist to disseminate her findings could well be overridden. Doubtless, in relation to most academic research such contingencies are exceptions, and should be treated as such. Nevertheless, given the high risk to human life and health posed by misuse of research in synthetic biology and related areas, such biological research constitutes a special case. Censorship of academic research needs special justification. However, that justification is, in general terms, available in the areas in question, e.g., the high risk of misuse by terrorists of such research. Naturally, censorship of any specific research or research project will not only need some justification, it will need a *specific* justification that details the high risk of misuse of this specific research project outcome by terrorists, e.g., the research outcome is a highly virulent, easily transmissible and readily weaponised pathogen.

Progress in Science

Though the above objection to censorship on the basis of academic freedom is legitimate it is not necessarily decisive in all cases. However, there are other arguments typically put against censorship of a more utilitarian kind. Chief among these is the argument from scientific progress: academic freedom, including freedom of dissemination, is necessary for scientific progress.

Things like "scientific openness", the "free sharing of information", and "academic freedom" are a matter of degree. As such they should be recognized as ideals rather than accurate characterizations or requirements of science. The perfect realization of such things cannot be essential to the progress of science, because science has already progressed, and continues to progress with great rapidity, despite the fact that they have never been perfectly realized. On the other hand, this does not show that the substantial, if imperfect, realisation of these ideals is not necessary for the progress of science.

Apparent counter-examples to the idea that scientific openness and the free sharing of information are essential to the progress of science are provided by nuclear physics (which has involved a long history of censorship) and the keeping of trade secrets in science-based industries. On the other hand, it could be argued that progress in nuclear physics depended ultimately on the freely disseminated work of Einstein and others.

Similar points can be made about "academic freedom" and "freedom of inquiry" more generally. Given that researchers' choices are so heavily limited by things like the availability of funding and other resources, and government and industry driven imperatives, for example, the idea that such freedoms are essential to science and its progress is, arguably, to some extent undermined by the status quo. Science moves forward despite the reality of various kinds of limits to the freedoms enumerated above. However, once again, the existence of such limits does not demonstrate that academic freedom and freedom of inquiry are not necessary for long term progress in science, much less that these things are not good in themselves.

One might admit that science has progressed despite these admittedly real constraints on freedom but nonetheless claim that science would have advanced *even further than* it has if there had been more freedom and openness in science than has actually been the case. Even if correct, however, this would not go to show that no restrictions on the dissemination of scientific information are warranted. The progress of science is just one of many legitimate social aims that must be taken into consideration by scientists and policy makers alike. The progress of science is important—as is the human right to freedom of inquiry and the institutional right to academic freedom—but other things such as public health/security are important too; and there is no compelling reason to think that these two kinds of goals will never conflict or that the former should always be given absolute priority over the latter (or vice-versa), in cases of conflict, regardless of the extent to which the latter is threatened.

A commonsense position is that trade-offs need to be made between, say, rights to disseminate and scientific progress on the one hand, and security/public health needs on the other, and that a reasonable balance must be struck between these and other values. On this view, we should sometimes be willing to make at least small sacrifices in the way of public health and/or security when this is necessary to achieve enormous benefits with regard to the progress of science; and we should sometimes be willing to make at least very small sacrifices with regard to the progress of science when this is necessary to achieve enormous benefits regarding public health and/or security. Either way, it should be recognized that, though there may be cases of conflict, the promotion of public health/security will often be instrumental in the promotion of science and that the promotion of science will often be instrumental in the promotion of public health and security. Public health and social stability are required for science to function, and the products of science will include means for protecting health and means for protecting society against potential adversaries. As we saw above, a more sophisticated view is that we can avoid at least some of these trade-offs by finding creative solutions that allow us to, for example, provide security but at little or no cost to academic freedom. Perhaps we can design a regulatory system that is creative in this sense (see next chapter).

Statement on Scientific Publication and Security

The idea that a balance must be struck between the promotion of security on the one hand and scientific freedom/openness on the other is expressed in the important joint "Statement on Scientific Publication and Security" of the "Journal Editors and Authors Group" which was simultaneously published by *Science*, *Nature*, the *Proceedings of the National Academy of Sciences*, and the American Society for Microbiology Journals in February 2003. This statement resulted from discussions between scientists, the scientific publishing community, and the security community at a workshop (motivated by public outcry over the mousepox and polio studies in particular) convened by the US National Academies of Science and the Center for Strategic and International Studies in January 2003. The statement says that:

> FIRST: The scientific information published in peer-reviewed research journals carries special status and confers unique responsibilities on editors and authors. We must protect the integrity of the scientific process by publishing manuscripts of high quality, in sufficient detail to permit reproducibility. Without independent verification, a requirement for

> scientific progress, we can neither advance biomedical research nor provide the knowledge base for building strong defense systems.
>
> SECOND: We recognize that the prospect of bioterrorism has raised legitimate concerns about the potential abuse of published information, but also recognize that the research in the very same fields will be critical to society in meeting the challenges of defense. We are committed to dealing responsibly and effectively with safety and security issues that may be raised by papers submitted for publication, and to increasing our capacity to identify such issues as they arise
>
> THIRD: Scientists and their journals should consider the appropriate level and design of processes to accomplish effective review of papers that raise such security issues. Journals in disciplines that have attracted numbers of such papers have already devised procedures that might be employed as models in considering process design. Some of us represent some of those journals; others among us are committed to the timely implementation of such processes, about which we will notify our readers and authors.
>
> FOURTH: We recognize that on occasion an editor may conclude that the potential harm of publication outweighs the potential societal benefits. Under such circumstances, the paper should be modified or not be published. Scientific information is also communicated by other means: seminars, meetings, electronic posting, etc. Journals and scientific societies can play an important role in encouraging investigators to communicate results of research in ways that maximize public benefits and minimize risks of misuse [28].

The fourth point concludes that there are at least imaginable situations where the censorship of science would be warranted. For an example of such a case, imagine a scenario where a researcher (accidentally or otherwise) finds an easy way to make a microbe as contagious, deadly, and untreatable as smallpox [59]. Release of this kind of information into the public domain could in no time lead to a global catastrophe of enormous proportions assuming there are those in the world who would use the discovery for malevolent purposes. A prudent response to a case like this would be to limit dissemination of the knowledge in question at least until such time as a vaccine or treatment for the microbe was developed and made widely available [59]. Developing and producing such defenses would of course require sharing the relevant information with *some* scientists—but those with the requisite expertise could be informed in confidence without disclosure to the public at large via publication. (Here and elsewhere it is crucial to keep in mind the variety of ways other than publication that information may be disseminated.) In the meanwhile it would be exceedingly risky to base the well-being of global humanity on the assumption that there is unlikely to be any person or group in the world with the capacity, opportunity and motivation to use such a discovery for malevolent purposes. That would be a highly dubious assumption; and the consequences of relying on it could be apocalyptic.

Why would anyone favour publication in this or other cases where it is (by hypothesis) determined that disclosure via publication would do immensely more harm than good all things considered? Questions about *when* publication would do more harm than good are of course partly empirical matters. They are also partly ethical matters, e.g., What are the greatest harms—loss of life or loss of freedom? It might be safe to assume that publication and the sharing of information usually do more good than harm; but one would be hard pressed to provide an *a priori* argument that publication is always in principle to be preferred to censorship unless one implausibly assumes that openness in science and the free sharing of information are the most important goods that there are. The idea that publication should be considered appropriate even when publication would do much more harm than good, when the harms in question are substantial loss of life, and when the prospect of harm is a clear and present danger is, frankly, indefensible.

Critics such as Ian Ramshaw, however, claim that it is too late for censorship now. There is already more than plenty of dangerous biological information out there for anyone who wants to cause major devastation, according to Ramshaw, so implementation of censorship now could not stop someone that was determined to use biological know-how to cause great harm. The fact that there is already lots of dangerous information in the published literature, if correct, however, does not necessarily imply that we should not pay attention to the dangers of future publications; and perhaps nothing as dangerous as our imaginary discovery has already been published. The weapons implications of previously published studies, in any case, will not always be obvious to would-be bioterrorists. Many would argue that publication of a book which synthesised the kinds of already published studies that Ramshaw has in mind, and explicitly explained the weapons implications, would be ill-advised. It is one thing if the information is out there but buried in the literature. And it would be another thing to explicitly uncover and advertise the weapons potential of existing "public knowledge"—at least if we are not yet prepared to protect ourselves against the potential weapons in question. Some may argue, on the other hand, that we need to publish such a book in order to better alert the scientific community to the kinds of things we need to prepare to protect ourselves against.

While the above "Statement on Scientific Publication and Security" is, in any case, considered by many to be a reasonable and responsible (scientific and publishing community) response to the dual-use dilemma, hard questions remain regarding (1) how the risks and benefits of publication should in practice be determined in any given case, and (2) who should have ultimate authority to make decisions about the censorship of science. The Statement claims that editors may sometimes make censorship decisions, but it gives no reasons for thinking that editors, or the scientific community in general, for that matter, are especially qualified to judge security risks. An important question thus concerns the extent to which the government and/or security community should be involved in scientific censorship.

In the US, at least, the government can only classify research findings when it has funded the relevant research. The only two exceptions involve cases of nuclear science discoveries with weapons implications and information related to patent applications. Aside from classification, the US government may choose to restrict the flow of information deemed to be "sensitive", e.g., in virtue of the security threat involved. The scientific community has, however, strongly resisted suggestions that the categorization of "sensitive" information should be used as a means of censoring science. In addition to concerns about government gaining too much control over the scientific enterprise is the worry that, in comparison with information that is "classified", the category of "sensitive" information is too vague. One of several worries is that scientists would be deterred from working or publishing on subjects that might (arbitrarily and unpredictably) get deemed sensitive and that important areas of research would be stalled as a result.

In an important (2004) report titled *Biotechnology Research in an Age of Terrorism*, widely known as "the Fink Report", the US National Research Council (NRC) recommends that, rather than government control over publication decisions in science, we should rely on voluntary self-governance of the scientific community in matters of censorship. The American Medical Association (AMA) recently adopted a set of "Guidelines to Prevent the Malevolent Use of Biomedical Research" that expresses similar sentiments. The AMA's Council on Ethical and Judicial Affairs claims that "physician-researchers, who possess profound knowledge of their research and of human health and disease, are arguably in the best position to assess the potential for and the ramifications of misapplications of their research" [24]. In addition to other worries about the governmental control over science, we here have a claim that scientists are themselves generally best qualified to assess the risks and benefits of publication.

There are important differences between American and Australian law with regard to the possible classification of information by the government. Most importantly, in Australia research need not be funded by government in order for the government to make research findings classified. There is thus more scope for the Australian government to classify information without resorting to the concept of "sensitive" information (which is not to say that some of the above-mentioned concerns about control over sensitive information would not apply to classification of information in the Australian context).

Moreover, there are important distinctions to be made in relation to dual-use publications. Some dual-use publications in the scientific literature are more directly applicable to harmful purposes than others. The demonstration that a mousepox virus can be engineered for increased virulence can be directly applied to attempt enhancement of smallpox virulence, indeed exactly the same technologies described in the mousepox publication could be used. The publication of the sequence for the 1918 influenza strain could be used to easily reconstruct this virus for harmful intent— the 2005 1918 flu "resurrection" study could simply be replicated. Other published discoveries posing dual-use dilemmas, however, cannot be used for malevolent purposes until significant further research is done. The demonstration that virus genomes can be engineered to encode antibodies that suppress immune responses could be ultimately used for harmful intent. However, a significant amount of research would need to be undertaken before a disseminating infectious agent could be constructed. The original use of this technology was to prevent graft rejection. Accordingly, we ought to develop a taxonomy of dual-use publications which at the very least should distinguish between 1st tier and 2nd tier research; the former refers to dual-use research possessed of direct applicability for harmful intent and the latter only with indirect applicability for harmful intent. (Strictly speaking, of course, directness of applicability will be a matter of degree.) First tier research findings might need to be censored or presented for publication in a manner that would not enable readers to replicate the experiments in question and thereby generate dangerous pathogens and the like. (Limited dissemination should, of course, be permitted to the extent necessary for purposes of replication/verification by those with adequate security clearance.) While omission of (detailed description of) materials and methods from published articles will sometimes be sufficient to prevent malevolent use of dual-use discoveries, this will not always be the case. There will be many situations where the general idea of what was discovered will be critical while the materials and methods would be obvious to anyone "skilled in the art".

Private Sector Research

Though much of what we have said above has explicitly focused on university research in particular, it is important to highlight the extent to which private sector research is also relevant to—and raises special issues for—the dual-use dilemma. The biotechnology industry has, of course, been a key driver of the genetic sciences which have made the dual-use dilemma so prominent in recent years. While the free sharing of information and the pursuit of knowledge are central values of academic research, however, industrial research is usually primarily motivated by the pursuit of profits. This has several implications. One, as we have pointed out before, is that industrial discoveries are often kept secret rather than published or otherwise widely shared with the academic community. While this might alleviate worries about the publication of dangerous discoveries that might result from such research (and, as we indicated before, make one doubt that complete openness in the sharing of information is essential to the progress of science), it also means that there is less public awareness of the discoveries taking place—and less public scrutiny of the extent to which the relevant research is dangerous. While academic scientists presumably usually have the good of humanity as a top concern, furthermore, one might be justified to worry that private companies will sometimes be comparatively less reluctant to pursue research with the potential for malevolent use in so far as their primary concern is (by admission) the promotion of profits rather than, necessarily, the benefit of humankind. A final cause for concern, finally, is that private research is not generally subject to the same institutional oversight (via institutional ethics committees) as that which takes place in universities. We point all this out to show that research of private industry is relevant to the dual-use dilemma too—and that it raises special issues of its own. Free enterprise is an important social good; but, because society may be threatened by dangerous industrial research, it may be argued that more oversight of industrial research is warranted. For example, companies that commercially synthesise the genes of viruses, e.g., potentially Ebola and smallpox, should be required to report any orders for genes of infectious organisms that could be used as weapons of mass destruction. In the western world this would be feasible. However, much of this work is now carried out in China and India where regulations to this effect would be far more difficult to impose and enforce. These points should be kept in mind in the discussion of regulation that follows.

Chapter 5
Options for the Regulation of Dual-Use Experiments and Information

In what follows we provide a set of options regarding the imposition of limits on dual-use experiments and the dissemination of potentially dangerous information resulting from dual-use discoveries. The five options presented range from the least intrusive/restrictive to the most intrusive/restrictive, and the advantages and disadvantages of each option are discussed in turn. Each option is complex in that it consists of six sub-options each of which pertains to one of the following six categories of decision.

Permissible or Impermissible Research

Who is to be the decision-maker in relation to determining whether or not an instance of one of the eleven identified types of experiment of concern is permissible or impermissible? The candidates for decision-maker are: the individual researcher; the specific institution hosting the research project in question, i.e., a university, corporation or government research centre; an independent authority; the government. In the case of the university, the decision-maker would presumably be a collegial body comprised of relevant scientists (at least).

Freedom of inquiry is a human right that finds institutional expression in universities in the form of academic freedom. In the context of a liberal democracy there is a presumption against governmental restriction of human rights, including in the name of protecting other human rights. Moreover, arguably progress in science is importantly dependent on academic freedom. Thus the development of countermeasures to existing bioterrorist threats might rely on a high degree of scientific freedom. As we have seen above, the question of whether research is morally permissible or impermissible is an extremely difficult issue, and it is by no means obvious who the ultimate decision-maker ought to be.

S. Miller, M.J. Selgelid, *Ethical and Philosophical Consideration of the Dual-Use Dilemma in the Biological Sciences*,

Mandatory Physical Safety and Security Regulation

Should there be regulations providing for mandatory physical safety and security of the storage, transport and physical access to samples of pathogens, equipment, laboratories etc.? The answer is presumably in the affirmative.

In theory, the specific content of these regulations might be determined either by a government agency, an independent authority, a professional association of scientists or the specific institution hosting the research programs in question. However, governments bear the ultimate institutional and moral responsibility for the safety and security of their citizens, including the researchers themselves, in so far as that safety and security is a matter of the physical conditions under which potentially harmful (albeit permissible) research is to be undertaken and the physical elements thereof stored, transported etc. Accordingly, the government would at least need to be able to satisfy itself that the regulatory system, including the regulations and their enforcement mechanisms, governing the physical safety and security of dual-use experimentation are adequate.

The application of many of these regulations could be undertaken by, for example, biosafety committees operating at the institutional level, e.g., a university-based biosafety committee. However, these committees would need in turn to be accountable to government (perhaps via an independent authority).

Licensing of Dual-Use Technologies/Techniques

Should there be mandatory licensing of dual-use technologies/techniques/DNA synthesisers/pathogen samples? Only certain laboratories in the public sector and the private sector might be licensed to engage in research involving the use of certain dual-use technologies. For example, laboratories that undertake genetic engineering of pox viruses would have the means to make recombinant smallpox viruses. Similarly, researchers studying influenza virus and using reverse genetics could easily construct a 1918 flu virus with available information.

The establishment of a licensing authority to conduct such a licensing process would be a significant addition to the mechanisms available to contain the dangers associated with the dual-use dilemma. However, it raises a number of important questions. One set of questions concerns the criteria that the licensing authority would deploy in its licensing process. Is there a presumption in favour of granting a licence; the criteria having been framed for the sole purpose of eliminating licence applicants that are manifestly unable to provide a safe and secure research environment? Are the criteria to be used to determine the issuing of licences objective and publicly available? Another set of questions pertains to the status and make-up of the licensing authority: is it, for example, independent of government in the sense that its decisions are binding and not able to be overridden by government?

These questions are important in the context of the concerns one might have in relation to government interference with freedom of intellectual inquiry (who decides what is permissible research?) and freedom of speech/dissemination of research findings (who decides what research findings can be disseminated and to whom?). The point is that a licensing authority could be given, at least in principle, powers that would in effect override human rights to freedom of intellectual inquiry and freedom of dissemination (and associated rights to academic freedom) by licensing, say, only government research centres.

Mandatory Education and Training

Given the potential harms arising from the eleven identified types of experiments of concern it is clear that some process of education and/or training for relevant researchers and other personnel is called for. There is a question as to the precise content of such education and training. However, at the very least those working in laboratories would need to have received safety and security training in relation to the physical safety and security of the storage, transport and physical access to samples of pathogens, equipment, laboratories etc. In addition, there is a need to ensure that editors and others responsible for the dissemination of potentially harmful information are aware of the issues in relation to dual-use research findings. In short, some forms of mandatory education and/or training are justified. What the precise content of such education/training programs ought to be, and who ought to be responsible for their provision, remain open questions. However, it is an institutional and moral responsibility of government to ensure that minimal training/education programs in relation to potentially harmful dual-use research and dissemination of dual-use research findings are being provided (even if not by government itself).

Mandatory Personnel Security Regulation

Physical safety and security of a research environment, including access by non-authorised persons, e.g., potential thieves, is one thing; however, personnel security in relation to researchers, e.g., background checks, screening of researchers in relation to any history of mental illness, political affiliations with extremist groups etc., is quite another. Doubtless it is prudent, indeed it is a moral requirement, that access to virulent pathogens be disallowed to a researcher diagnosed as a psychopath or to a known member of a terrorist organisation. On the other hand, other things being equal, government officials prying into the lives of university students enrolled in degrees in the biological sciences is an unwarranted intrusion of civil liberties. Here, as elsewhere, the devil is in the detail, and there is a need for specific policies to be framed in the light of a range of human rights, academic and scientific considerations as well as security concerns.

One way forward here might be to develop a system of security checks for personnel working in licensed laboratories but (absent special considerations) not for other research personnel.

Censorship/Constraint of Dissemination

As we have seen above, the question of whether research findings ought to be freely disseminated, censored or their dissemination in some lesser way restricted is an extremely difficult issue and it is by no means obvious who the ultimate decision-maker ought to be. Freedom of speech and freedom of dissemination of knowledge are human rights that find institutional expression in universities in the form of academic freedom. In the context of a liberal democracy there is a presumption against governmental restriction of human rights, including in the name of protecting other human rights. Moreover, arguably progress in science is importantly dependent on scientific—and, therefore, academic—freedom. On the other hand, restrictions on the dissemination of new scientific research that is likely to facilitate the malevolent purposes of bioterrorists is warranted.

A relevant important distinction here is that made above between 1st tier and 2nd tier dual-use research. For example, 1st tier research findings might need to be disseminated in such a way that anyone being informed of these findings would not be able to replicate the experiments that enabled the results reported in the findings. (It goes without saying, nevertheless, that experiments would need to be replicated (by those with adequate security clearance) to the extent necessary to ensure scientific verification.)

A final point here is that given the status attached to numbers and quality of publications by scientists—and the corresponding connection between status seeking behaviour and scientific advancement—censoring or otherwise restricting the dissemination of scientific work may to a corresponding extent undermine scientific advancement, unless alternative reward structures for the scientists in question are developed, e.g., monetary payments, medals.

Option 1—Complete Autonomy of the Individual Scientist

The least intrusive/restrictive option is, of course, to do nothing about the dual-use dilemma. This laissez-faire option would allow the scientific status quo to run its own course. Individual researchers would be left to make their own decisions whether to engage in particular experiments of concern, and individual researchers and editors would be left to make their own decisions about whether or not to disseminate the findings of that research. Moreover, it would be up to individual researchers, editors and/or laboratories to determine safety and security protocols and to educate themselves about the dual-use dilemma and the potential dangers of publication and other forms of information dissemination. Scientists would be free to present whatever they consider to be appropriate at conferences, and free to discuss whatever they consider to be appropriate with colleagues.

Advantages

Option 1 maximally promotes the autonomy of the individual researcher and editor, and of the scientific community; it manifests an extremely high level of trust in individual scientists and the scientific community as a whole. It acknowledges that scientists as individuals and the scientific community as a whole are generally responsible, that they generally aim at community benefit, and it assumes that they embody sufficient expertise to assess risks and make correct decisions about when a particular experiment of concern or form of dissemination is appropriate or not. It is possible that the flourishing of science would be maximally promoted via this option (assuming that security is not overly compromised in the process—as science would be compromised in the event of a security breakdown).

Disadvantages

Experience shows that individual scientists and the scientific community as a whole cannot always be relied upon to govern themselves in a responsible manner. A long history of unethical research on human subjects (e.g., research by the Nazis during World War II and the Tuskegee Syphilis Study conducted by the US Public Health Service between 1932 and 1972) and other examples of irresponsible conduct of scientists (e.g., involving conflicts of interest and fraud) provide counterexamples to the idea that voluntary self-governance of scientists can be relied upon. It is for this reason that so many research ethics guidelines (having various degrees of legislative force) and requirements regarding ethics education for scientists have been put into place worldwide. Even if the vast majority of scientists have the best of intentions, it cannot be assumed that *every* scientist will do the right thing if left to his or her own devices.

In addition to questioning the prudence of placing so much trust in the moral rectitude of (all) scientists, some will challenge the idea that the scientific community has sufficient expertise for addressing dual-use risks. Many (perhaps most) life scientists are not well aware of things like the dual-use phenomenon and specific details of the Biological Weapons Convention (BWC)—and many (perhaps most) are not well aware of the specific ways in which their own discoveries might be misused by those with malevolent intentions. While it is true that scientists who are well aware of the dual-use phenomenon may sometimes be best able to recognize the scientific potential of a discovery to be used for things like biological weapons-making, scientists are generally not security experts. The latter point is that there is no reason to think that scientists have sufficient expertise to judge the security implications of the malevolent use of their discoveries (even if they are well able to recognize the *scientific potential* of malevolent use).

Moreover, there are situations where scientists themselves are systematically denied information crucial to the assessment of the risks and benefits of their work [58, 59]. The danger of publication in the case of the mousepox study (which, has been a central focus of debate over the censorship of science in the context of the bioterrorist threat related to dual-use discoveries) largely depends on the likelihood of smallpox proliferation insofar as would-be bioterrorists would require access to the smallpox virus in order to apply the mousepox genetic engineering technique to it (with the aim of producing vaccine-resistant smallpox). Any specific details about the likelihood of smallpox proliferation, from former Soviet weapons stocks of the virus for example, are classified information that scientists simply lack access to. The scientists and editors involved with publication of the mousepox study presumably thus lacked information which was arguably crucial to assessing the security risks of publishing that study. Microbiologists may have been best able to judge the likelihood that the engineering technique employed with mousepox would enable production of vaccine-resistant smallpox if applied to the smallpox virus; and public health experts would be best able to assess the public health impact that would result from an attack involving such a virus. Neither group of scientists, however, would have particular expertise for assessing (1) the likelihood that bioterrorists would be able to access the smallpox virus to begin with, (2) the likelihood that such terrorists would have the means to apply the mousepox technique to the smallpox virus, (3) the likely existence of the intention such terrorists might have to engineer and use such a virus, or (4) the security (as opposed to public health) impact of an attack involving such a virus. The suggestion that scientists (or editors) should be left to act on their own assessments of whether or not overall benefits outweigh risks of publication (or other forms of information dissemination) in any given situation is thus open to challenge.

Further, Option 1 does nothing to address the issue of the tendency to bias and, in some cases, conflicts of interest that inevitably face scientists. Given the reward system in academia, scientists are under pressure to publish as much of their work as possible. The aim to advance one's own career may thus conflict with one's aim to protect security (by voluntarily suppressing publication of one's own work). A related point is that it is not unlikely that the value system of the typical scientist will place high (and arguably disproportionate) weight on the value of scientific progress and freedom of inquiry vis-à-vis the importance of security [59]. In the private sector, on the other hand, high (and arguably disproportionate) weight is likely to be placed on the promotion of profits and free-enterprise.

Finally, Option 1 fails to accord to the public and the government in particular its moral right—and in the case of government, its moral responsibility—to safeguard against (at the very least) clear and present dangers to health (including life). A clear and present danger in the form of, say, an artificially created pandemic outweighs any moral right to freedom of inquiry and dissemination scientists might have; roughly speaking, the right to life in such situations outweighs the right to freedom. Moreover, the government has a moral responsibility to act to protect its citizenry in such situations and, more generally, a responsibility to ensure reasonable preventative measures against such threats. Accordingly, there is a moral requirement for regulatory systems accountable to government; such systems would at the very least require minimal physical safety and security and mandatory education/training in relation to physical safety and security.

Because its solution to the problem of the dual-use dilemma is in effect to do nothing about it, Option 1 is unlikely to be considered by many to be a morally responsible course to take. We endorse this view of Option 1 and recommend against its adoption.

Option 2—Institutional Control

Recognising the inadequacy of Option 1 to address legitimate security concerns but acknowledging, nevertheless, the importance of freedom of intellectual inquiry and dissemination—and especially academic freedom—and of progress in science, Option 2 opts for: (i) government regulation (applied within institutions, i.e., universities, corporations and government research centres) in relation to physical safety and security (roughly speaking, the conditions under which dual-use research is conducted) and (ii) collegial decision-making in relation to the permissibility of dual-use research (at least in the case of university based research). Unregulated dual-use research—e.g., by non-institutionally based researchers—would be outlawed. However, there would be no licensing of dual-use technologies or mandatory personnel security provisions. Moreover, dissemination of dual-use findings of university based researchers would be a matter for individual researchers and editors to decide. The dissemination of the dual-use findings of researchers hosted by corporations and government research centres would be a matter for determination by those institutions.

The principal institutional mechanism for determining the permissibility of dual-use research would be institutional biosafety committees (IBCs). For example, university-based IBCs would be expected to apply government regulations in university based research centres. Under this arrangement researchers would be required to submit any research proposals falling within categories of concern to IBCs for review. In cases where sufficient dangers of experimentation are foreseen, the IBCs would make an adjudication that the research project in question not go ahead. In relation to dual-use research, these institutional committees would be the ultimate decision-makers (at least in universities, though presumably not in corporations and government research centres). Other things being equal governments, for example, would not be able to override the determinations of these IBCs. Naturally, other things might not be equal. For example, the decisions of the IBCs would have to comply with regulations enacted by government; to this extent they would be accountable to government.

Moreover, under Option 2 there would be some emphasis on education and training, albeit education and training would not be mandatory. For example, governments would be encouraged to become actively engaged in the promotion of education of scientists with regard to the dual-use dilemma and thus potential dangers of publication and other forms of information dissemination. This option would involve a campaign aimed at increasing awareness. Various means would be used to alert scientists to the general dangers highlighted in this report, and the specific dangers associated with particular areas of research—especially with regard to experiments of concern enumerated above. Reports would be published and disseminated by relevant research councils and governmental departments, government-sponsored seminars and/or short courses would be offered at relevant venues; scientist training programs, research institutes, and private companies would be advised about the dangers of dual-use, the importance of educating scientists about it, and resources (of various forms) would be offered to facilitate provision of appropriate education.

Advantages

A principal advantage of Option 2 is that unlike Option 1 it acknowledges security concerns and seeks to do something to address them, namely, regulate the research environment in relation to safety and security, and increase awareness of the dual-use dilemma. A further advantage is that it preserves most of the benefits of Option 1 in relation to academic freedom and progress in science.

The advantage of voluntary educational promotion, in particular, is that it offers much to—but demands little from—scientists, the scientific community, or industry. It thus aims to maximally promote scientific autonomy and self-regulation as in Option 1; but, by actively promoting education, it addresses important disadvantages of Option 1 as discussed above—i.e., specifically with regard to the (otherwise) general lack of scientists' awareness regarding the dual-use dilemma and specific dangers associated with their own research in particular. It expresses trust in scientists, the scientific community, and industry to do the right thing; but it assists them to do the right thing.

Disadvantages

The most important disadvantage of Option 2 is that it fails to *adequately* address security concerns. Arguably, without security provisions in relation to personnel and licensing of laboratories all manner of dangerous technologies, equipment and samples of pathogens will be made available to malevolent state and non-state actors. The regulatory system provided for in Option 2 is simply not adequate to the task at hand.

Given pressures on resources including time, it cannot be assumed that all relevant scientific institutions (including companies) would offer recommended education, and it cannot be assumed that all scientists would take advantage of such education, if educational requirements are not put into place. If specific content is not required, then quality of education may be compromised. If education is required, then a small degree of scientific autonomy would be lost, but this may only be considered a minor intrusion on the scientific community.

Education and training programs, especially voluntary education and training programs, will not turn scientists into security experts. Moreover, such education (presumably) could not include disclosure of classified information sometimes crucial to risk assessment. The ability of even the better educated scientists (that would be expected under Option 2) to assess security risks and benefits would thus remain limited as discussed in disadvantages of Option 1.

Finally, problems related to bias and conflicts of interest will remain. Pressure on scientists to publish their work will continue and, absent restrictions on dissemination of information, scientists would be at liberty to give in to this pressure.

Option 3—A Dual System: Institutional and Governmental Control

Acknowledging that Option 2 does not do enough to address security concerns, Option 3 provides for a greatly enhanced regulatory system. Specifically, Option 3 provides for mandatory personnel security, licensing of dual-use technologies and mandatory education and training (in addition to mandatory physical safety and security). This regulatory system would apply to both public and private sector research centres.

Notwithstanding this enhanced regulatory system, academic freedom and progress in science remain. As with Option 2, Option 3 would involve collegial decision-making by means of IBCs in relation to the permissibility of dual-use research (at least in the case of university based research).

Moreover, dissemination of dual-use findings of university based researchers would be a matter for individual researchers and editors. Dual-use research in, and the dissemination of the dual-use findings of researchers hosted by, corporations and government research centres would be a matter for determination by those institutions.

As with Option 2, in Option 3 the principal institutional mechanism for determining not only the permissibility of dual-use research, but also for applying the regulatory system in respect of physical safety and security would be the IBCs. However, on Option 3 the IBCs would have additional duties in relation to mandatory education and training, and the compliance of laboratories with the requirements of a licensing authority (including those pertaining to personnel security).

Option 3 involves the establishment of such an authority, albeit one independent of government. (See above for discussion of this issue.)

As indicated above, Option 3 provides an alternative to merely promoting education about the dual-use dilemma, namely, the formal requirement of this kind of training for those working in relevant fields of study (and relevant industries). Research ethics education is increasingly being required for those involved with research involving human or animal subjects worldwide (especially those receiving government funding). An educational requirement regarding the dual-use dilemma would fit in with this already existing trend. Specifications would be made about content to be covered in such training and heavy emphasis would be placed on experiments of concern in particular. Specific content most relevant to particular areas of research could be part of the educational requirement of those working in such areas. Given the grey area that exists between offensive and defensive biological weapons research, for example, it is especially important that those working in defence become especially familiar with the BWC and related issues. Those working with pox viruses would study the history of smallpox (which is important for understanding weapons threat of smallpox), and so on.

Advantages

By comparison with Option 2 (and Option 1), Option 3 gives a high priority to security concerns in relation to dual-use research; Option 3 offers a greatly enhanced regulatory system. Moreover, as was the case with Option 2, Option 3 substantially preserves academic freedom and the conditions conducive to progress in science.

By providing for mandatory (quality-controlled) dual-use education and training Option 3 goes beyond Option 2 and would, thereby, make scientists better aware of the potential dangers of experimentation and information dissemination. Moreover, researchers and other personnel would receive mandatory training in relation to key aspects of safety and security; this is surely a step in the right direction.

Disadvantages

Whether or not required, educational training would not turn scientists into security experts and such education (presumably) could not include disclosure of classified information sometimes crucial to risk assessment. The ability of even better educated scientists to assess security risks and benefits would thus remain limited as discussed in the disadvantages of Option 2.

Moreover, as with Option 2, problems related to bias and conflicts of interest will remain. Pressure on scientists to publish their work will continue and, absent restrictions on dissemination of information, scientists would be at liberty to give in to this pressure.

Option 4—An Independent Authority

Option 4 involves the establishment of an authority that is independent of both the research institutions (universities, corporations and government research centres) and government. This independent authority would be comprised of scientists, security experts (including those with the highest feasible level security clearance) and ethicists.

This independent authority would have ultimate decision-making powers in relation to both the conduct of dual-use research and the dissemination of dual-use research findings. Moreover, it might also constitute the above-mentioned independent authority issuing licences to laboratories in relation to dual-use technologies (and providing for personnel security). Other things being equal, decisions of this independent authority would not be able to be overridden by government. Naturally, other things might not be equal. For example, the decisions of the independent authority would have to comply with regulations enacted by government; to this extent it would be accountable to government.

In addition, this independent authority might have an accountability role on behalf of government in relation to the application of government regulations in respect of physical safety and security, and dual-use education and training. Its determinations in these respects would not be final; rather these determinations would have the status of advice to government.

Under this arrangement researchers might be required to submit any research proposals falling within categories of concern to IBCs for review; and they would be required to submit any research findings which end up falling within categories of concern to IBCs after the fact (i.e., if a relevant dual-use discovery is unexpectedly made). In cases where sufficient dangers of experimentation or of information dissemination are foreseen, the IBCs will issue prohibitions or refer the studies to the independent authority for determination.

It is important to note that even under this form of meta-regulation the independent authority would have the power to intervene at any lower level, including overturning decisions at the lower level and auditing the work of the IBCs. In effect, this independent body would have the ultimate authority to determine what was permissible or impermissible dual-use research, and to determine whether and in what form dual-use research findings could be disseminated.

In addition to the research screening process described above, a national code of scientific conduct including statements analogous to the American Medical Association's (AMA's) Guidelines to Prevent the Malevolent Use of Biomedical

Research would be developed by the independent authority. The code would include the requirement that scientists refer any research or research findings that falls within the eleven categories of experiments of concern to IBCs or other institutional ethics committees for determination. The code of conduct would be legally binding and apply to those working in industry as well as academia.

Advantages

Option 4 involves a decision making body embodying both the scientific and security expertise required for rigorous analysis of the security risks of research and of publication. Some would argue that this addresses an unacceptable deficiency of all previously enumerated options. On the other hand, under Options 2 and 3 security expertise/advice could be provided to the IBCs by government (within constraints regarding disclosure of classified information).

The code of conduct would help to ensure that any research falling within the category of types of experiments of concern will be sent for adjudication by those with more expertise than the (ordinary, educated) researcher. IBCs' and ethics committees' members will receive advanced training. The independent authority's members will have the highest degree of both scientific and security expertise. The lower level committees will, as necessary, refer especially difficult cases to the independent authority for final judgment. On the other hand, under Options 2 and 3 an enforceable code of conduct with similar benefits *could be* developed by the relevant professional association and applied by the IBCs.

Because ultimate decision making authority in problematic cases will not lie in the hands of either the individual researcher or the individual institution (or collegial committee of scientists, in the case of universities), previously mentioned problems regarding bias and conflicts of interest will be addressed. If the independent authority is appropriately constituted then it is less likely that its decisions would be biased towards either the promotion of science, freedom of inquiry/expression, or towards the promotion of security; perhaps a better balance between these aims is more likely to be achieved than is the case with Option 3 (and indeed Option 5—see below).

Finally, the two-tiered screening procedure would ensure a degree of efficiency: the independent authority would make determinations only in a fraction of cases, since IBCs would make determinations in the bulk of cases.

Disadvantages

Option 4 takes decision making authority out of the hands of individual researchers/editors and out of the hands of collegial committees of scientists, at least in the case of dual-use research and dissemination of the findings of dual-use research falling within categories of concern. Accordingly, by comparison with Options 1, 2 and 3, individual intellectual freedom, freedom of dissemination, and academic freedom are thereby diminished. Moreover, there may well be adverse effects on the progress of science. Also there is likely to be a cost to such a process and for the running of the independent body, and one would question the benefits of such a body in terms of efficiency and equity.

The possibility that publication would be prohibited in certain areas of potentially important research would deter scientists from working in such areas insofar as rewards and recognition associated with publication and other forms of dissemination (such as conference presentations) would be less certain. It is not unlikely that progress in areas of science important to the promotion of both human health and security/defence would be stalled as a result. Because it will not be determined if publication, for example, is permitted until after research takes place, researchers will avoid pursuit of projects where the likelihood of prohibition on publication (or other forms of dissemination) is greatest despite the importance of much of this research.

Option 5—Governmental Control

Option 5 would include a regulatory system authorised by, and accountable to, government that comprised mandatory physical safety and security, mandatory personnel security, and mandatory education and training. However, unlike Options 1, 2, 3, and 4 government would have ultimate and overriding decision-making authority in relation to both dual-use research and dual-use publication, and in relation to the licensing of laboratories.

Option 5 is consistent with government receiving *advice* from individual researchers/editors, non-government institutions, e.g., scientific associations, universities and corporations, and independent authorities.

Moreover Option 5 is consistent with various forms of meta-regulation, including a two-tiered system. For example, university-based biosafety committees (IBCs) might be expected to apply government regulations in university based research centres. Under this arrangement researchers might be required to submit any research proposals falling within categories of concern to IBCs for review; and they would be required to submit any research findings which end up falling within categories of concern to IBCs after the fact. In cases where sufficient dangers of experimentation or of information dissemination are foreseen, the IBCs will issue prohibitions or refer the studies to the relevant government committee for determination.

It is important to note that even under this form of meta-regulation the government committee would have the power to intervene at any lower level, including overturning decisions at the lower level and auditing the work of the IBCs. In effect, government would have the ultimate authority to determine what was permissible or impermissible dual-use research, and to determine whether and in what form dual-use research findings could be disseminated.

Advantages

Option 5 places decision making power over issues considered vital to security interests in the hands of the entity most responsible for national security—i.e., the government. In comparison with any previously enumerated options, Option 5 would furthermore place final decision making power in the hands of those with the highest levels of security expertise.

Moreover, a two-tiered screening procedure would ensure a degree of efficiency: the government committee makes determinations only in a fraction of cases, since IBCs would make determinations in the bulk of cases.

Disadvantages

In so far as scientific decision making is placed in the hands of government, a significant degree of academic (and industrial) freedom and scientific autonomy would be sacrificed. If scientists are required to forgo information dissemination, then the pursuit of, and progress in, important areas of research will be hindered. Accordingly, scientific progress would be reduced insofar as some discoveries would not become part of the general corpus of scientific knowledge.

There must be some doubt that a governmental body would in fact strike an appropriate balance between academic freedom and the progress of science, on the one hand, and the promotion of security in its decision making, on the other; for it is arguably in the nature of government to place a disproportionately high value on security in particular and, more generally, to seek to increase its sphere of control. Also the government may have a vested interest to ensure that dual-use research takes place but the scientific findings are secret. Some of the work under BioShield and Defense in the United States could fall into this category. It could lead to suggestion that the government had something to hide and was in fact involved in bioweapons research.

Some will go further and argue that governmental censorship of science threatens academic freedom, freedom of intellectual inquiry, freedom of speech, and ultimately the progress of science itself; the concern is that the proposed arrangement provides a slippery-slope down which there are ever increasing strictures on scientific research and dissemination.

In the light of the above-mentioned dangers attendant upon governmental control and the availability of other options, e.g., Options 3 and 4, that are able to address security concerns and enable governmental input, we recommend against the adoption of Option 5.

The five broad options and their sub-options can be represented in the form of a matrix, shown in Table 1.

Table 1 Decision-making for dual-use dilemmas in the biological sciences

	Options				
Decisions	Option 1—the complete autonomy of the individual scientist	Option 2—institutional control	Option 3—institutional & governmental control	Option 4—an independent authority	Option 5—governmental control
Who are the decision-makers regarding im/permissible research?	Individual resear-cher	(i) Scientists in university (collegial) (ii) Corporation (iii) Govt. Res. Centre	(i) Scientists in university (collegial) (ii) Corporation (iii) Govt. Res.Centre	Independent authority	Government
Should compliance with physical safety & security regulation be mandatory?	No	Yes	Yes	Yes	Yes
Should dual-use technology be licensed?	No	No	Yes	Yes	Yes
Should education & training be mandatory?	No	No	Yes	Yes	Yes
Should personnel security regulation be mandatory?	No	No	Yes	Yes	Yes
Who are the decision-makers regarding censorship/con-straint of material proposed for dissemination?	Individual editor	(i) Individual editor (ii) Corporation (iii) Govt. Res. Centre	(i) Individual editor (ii) Corporation (iii) Govt. Res. Centre	Independent authority	Government

NB: The decision-making in question pertains only to dual-use research in the biological sciences identified as potentially problematic by virtue of coming under one of the pre-established headings of Experiments of Concern

Conclusion

As we have seen, Options 1 and 5, namely the Complete Autonomy of the Individual Scientist and the Governmental Control options, exist at the two extremes of the spectrum of possibilities, and both have very significant, albeit contrasting, disadvantages. Option 1 is largely oblivious to the security threat; Option 5 gives insufficient weight, among other things, to the liberal democratic values of freedom of intellectual inquiry, freedom of communication and, relatedly, academic freedom. Moreover, the main advantages of Options 1 and 5 (autonomy and security, respectively) are in large part made available in one form or another by each of Options 2, 3 and 4. Accordingly, we have recommended against Options 1 and 5. What of the choice between Options 2, 3 and 4? Are there good and decisive reasons for preferring one of these over the others?

Recall that unlike Options 3 and 4, Option 2 (Institutional Control) does not involve mandatory licensing of technology, mandatory education/training or mandatory personnel security regulation. As argued above, these omissions are a weakness of Option 2. Moreover, Option 2 does not have any compensating advantages, particularly in comparison with Option 3. Accordingly, we recommend against Option 2.

Thus we are left with Option 3 (Institutional and Governmental Control) and Option 4 (An Independent Authority). Both options seem to us to be both feasible and ethically justifiable, depending on the precise institutional form that each might take.[6] Moreover, as things stand, we do not believe that there are good and decisive reasons in favour of one option over the other. Rather what is needed now is the development of a more detailed description of each of these two competing institutional models being proposed under Options 3 and 4 (respectively). Such a process of institutional design would enable a meaningful calibration of the advantages and disadvantages of the two competing models and, as a consequence, allow for the possibility of an informed decision to be made as to which to adopt.

[6] See also Miller [86, Chapter 6].

Bibliography

Journal Articles (* Published dual-use research findings)

1. Anonymous (2003). Statement on the consideration of biodefence and biosecurity. *Nature, 421*(6925), 771.
2. Atlas, R. M., & Reppy, J. (2005). Globalizing biosecurity. *Biosecurity and Bioterrorism, 3*(1), 51–60.
3. Badash, L. (2004). Science and social responsibility. *Minerva, 42*, 285–298.
4. Balmer, B. (2002). Killing without the distressing preliminaries: Scientists defence of the British biological warfare programme. *Minerva, 40*(1), 57–75.
5. Barletta, M., Sands, A., & Tucker, J. B. (2002). Keeping track of anthrax: The case for a biosecurity convention. *Bulletin of the Atomic Scientists, 58*(3), 57–62.
6. Beck, V. (2003). Advances in life sciences and bioterrorism. *EMBO Reports, 4*(special issue), S53–S56.
7. Block, S. M. (2001). The growing threat of biological weapons. *American Scientist, 89*(1), 28–38.
8. Block, S. M. (2002). A not-so-cheap stunt. *Science, 297*, 769.
9. *Cello, J., Paul, A. V., & Wimmer, E. (2002). Chemical synthesis of poliovirus cDNA: Generation of infectious virus in the absence of natural template. *Science, 297*(5583), 1016–1018.
10. Check, E. (2003). US army attacked over published patent for bioweapons grenade. *Nature, 423*(6942), 789.
11. Choffnes, E. (2002). Bioweapons: New labs, more terror? *Bulletin of the Atomic Scientists, 58*(5), 29–32.
12. Cook-Deegan, R. M., Berkelman, R., Davidson, E. M., Finder, S., Heitman, E., Kelley, M. C., King, N. M. P., Moseley, R., Thomas, J. C., Tilden, S. J., & Vangsnes, N. M. (2005). Issues in biosecurity and biosafety. *Science, 308*, 1867–1868.
13. Couzin, J. (2002). Active poliovirus baked from scratch. *Science, 297*, 174–175.
14. Couzin, J. (2004). US agencies unveil plan for biosecurity peer review. *Science, 303*(5664), 1595.
15. Cozzarelli, N. R. (2003). PNAS policy on publication of sensitive material in the life sciences. *Proceedings of the National Academy of Sciences, 100*(4), 1463.
16. Eckenwiler, L. (2005). Ethics and the underpinnings of policy in biodefense and emergency pre-paredness. *Cambridge Quarterly of Healthcare Ethics, 14*, 306–315.
17. Editorial (2003). Censoring science. *Canadian Medical Association Journal, 168*(12), 1517.
18. Ember, L. (2004). Securing dual-use biology. *Chemical and Engineering News, 82*(11), 11.
19. Enserink, M. (2004). Tiptoeing around pandoras box. *Science, 305*(5684), 594–595.
20. Enserink, M., & Kaiser, J. (2005). Has biodefense gone overboard? *Science, 307*(5714), 1396–1398.
21. Epstein, G. L. (2001). Controlling biological warfare threats: Resolving potential tensions among the research community, industry, and the national security community. *Critical Reviews in Microbiology, 27*(4), 321–354.

22. Falkow, S. (2003). "Statement on Scientific Publication and Security" fails to provide necessary guidelines. *Proceedings of the National Academy of Sciences, 100*(10), 5575–5576.
23. Frishknecht, F. (2003). The history of biological warfare. *EMBO Reports, 4*(special issue), S47–S52.
24. Green, S. K., Taub, S., Morin, K., & Higginson, D. (for the Counceil on Ethical, Judicial Affairs of the American Medical Association) (2006). Position paper: Guidelines to prevent malevolent use of biomedical research. *Cambridge Quarterly of Healthcare Ethics, 15*(4), 432–439.
25. Henderson, D. A., & Fenner, F. (2001). Recent events and observations pertaining to smallpox virus destruction in 2002. *Clinical Infectious Diseases, 33*(7), 1057–1059.
26. *Jackson, R. J., Ramsay, A. J., Christensen, C. D., Beaton, S., Hall, D. F., & Ramshaw, I. A. (2001). Expression of mouse interleukin-4 by a recombinant ectromelia virus suppresses cytolytic lymphocyte responses and overcomes genetic resistance to mousepox. *Journal of Virology, 75*(3), 1205–1210.
27. *Johnston, J. B., Nazarian, S. H., & McFadden, R. N. (2005). Myxoma virus infection of primary fibroblasts varies with cellular age and is regulated by host interferon responses. *Virology, 332*(1), 235–248.
28. Journal Editors, Authors Group (2003). Uncensored exchange of scientific results. *PNAS, 100*(4), 1464.
29. Kahn, L. H. (2003). Life scientists and the dual use dilemma. *Science, 302*(5652), 1894–1895.
30. Kaiser, J. (2005). Resurrected influenza virus yields secrets of deadly 1918 pandemic. *Science, 310*(5745), 28–29.
31. Keel, B. A. (2004). Protecting Americas secrets while maintaining academic freedom. *Academic Medicine, 79*(4), 333–342.
32. Kelley, M., & Coghlan, J. (2003). Mixing bugs and bombs. *Bulletin of the Atomic Scientists, 59*(5), 24–31.
33. Kelley, M. (2006). Infectious disease research and dual-use risk. *Virtual Mentor, 8*(4), 230–234.
34. Kennedy, D. (2002). Response to Steven M. Block's "A not-so-cheap stunt". *Science, 297*, 769–770.
35. Kennedy, D. (2003). Two cultures. *Science, 299*, 1148.
36. Kwik, G., Fitzgerald, J., Inglesby, T. V., & O'Toole, T. (2003). Biosecurity: Responsible stewardship of bioscience in an age of catastrophic terrorism. *Biosecurity and Bioterrorism, 1*(1), 27–35.
37. Lachmann, P. J. (2005). Microbial subversion of the immune response. *Proceedings of the National Academy of Sciences, 99*(13), 8461–8462.
38. Leitenberg, M. (2003). Distinguishing offensive from defensive biological weapons research. *Critical Reviews in Microbiology, 29*(3), 223–257.
39. Leitenberg, M., Leonard, J., & Spertzel, R. (2003). Biodefense crossing the line. *Politics and the Life Sciences, 22*(2), 2–3.
40. MacLeod, R. (2002). Preface. *Minerva, 40*(1), 1–2.
41. Müllbacher, A., & Lobigs, M. (2001). Creation of killer poxvirus could have been predicted. *Journal of Virology, 75*(18), 8353–8455.
42. Nixdorff, K., & Bender, W. (2002). Ethics of university research, biotechnology and potential military spin-off. *Minerva, 40*(1), 15–35.
43. Petro, J. B., & Relman, D. A. (2003). Understanding threats to scientific openness. *Science, 302*, 1898.
44. *Pomerantsev, A. P., Staritsin, N.A., Mockov, Yu. V., & Marinin, L. I. (1997). Expression of cereolysine AB genes in *Bacillus anthracis* vaccine strain ensures protection against experimental hemolytic anthrax infection. *Vaccine, 15*(17/18), 1846–1850.
45. Quester, G. H. (2000). Mismatched deterrents: preventing the use of nuclear, biological, and chemical weapons. *International Studies Perspectives, 1*, 165–176.

46. Rabodzey, A. (2003). Biosecurity implications of the synthesis of pathogenic viruses. *Politics and the Life Sciences, 22*(2), 44–49.
47. Racaniello, V. R., & Baltimore, D. (1981). Cloned poliovirus complementary DNA is infectious in mammalian cells. *Science, 214*, 916–919.
48. Rath, J., Bernhard, J., & Doblhoff-Dier, O. (2004). Journals must cooperate to defend biosecurity. *Nature, 432*(7017), 549.
49. *Read, T. D. et al. (2003). The genome sequence of *Bacillus anthracis* ames and comparison to closely related bacteria. *Nature, 423*(6935), 81–86.
50. Resnik, D. B. (2005). Conflicts of interest at the NIH: No easy solution. *Hastings Center Report, 35*(1), 18–20.
51. *Richman, D. D. (2006, preprint), Antiviral drug resistance. *Antiviral Research* Available at: http://www.elsevier.com/locate/antiviral.
52. *Robbins, S. J., Jackson, R. J., Fenner, F., Beaton, S., Medveczky, J., Ramshaw, I. A., & Ramsay, A. J. (2005). The efficacy of cidofovir treatment of mice infected with ectromelia (Mousepox) virus encoding interleukin-4. *Antiviral Research, 66*(1), 1–7.
53. *Rosengard, A. M., Liu, Y., Nie, Z., & Jimenez, R. (2002). Variola virus immune evasion design: Expression of a highly efficient inhibitor of human complement. *Proceedings of the National Academy of Science, 99*(13), 8808–8813.
54. Schulz, W (2003). Biotechnology research reviews. *Chemical and Engineering News, 81*(41), 15.
55. Schwellenbach, N. (2005). Biodefense: A plague of researchers. *Bulletin of the Atomic Scientists, 61*(3), 14–16.
56. Selgelid, M. J. (2003). Smallpox revisited? *The American Journal of Bioethics, 3*(1): W5–W11.
57. Selgelid, M. J. (2004). Bioterrorism and smallpox planning: Information and voluntary vaccination. *Journal of Medical Ethics, 30*, 558–560.
58. Selgelid, M. J. (2006). The ethics of dangerous discovery. *Cambridge Quarterly of Healthcare Ethics, 15*(4): 444–447.
59. Selgelid, M. J. (2007). A tale of two studies: Ethics, bioterrorism, and the censorship of science. *The Hastings Center Report, 37*(3), 35–43.
60. *Shimono, N., Morici, L., Casali, N., Cantrell, S., Sidders, B., Ehrt, S., & Riley, L. W. (2003). Hypervirulent mutant of mycobacterium tuberculosis resulting from disruption of the mce1 operon. *Proceedings of the National Academy of Sciences, 100*(26), 15918–15923.
61. Simon, J., & Hersh, M. (2002). An educational imperative: The role of ethical codes and normative prohibitions in CBW-applicable research. *Minerva, 40*(1), 37–55.
62. Somerville, M. A., & Atlas, R. M. (2005). Ethics: A weapon to counter bioterrorism. *Science, 307*(5717), 1881–1882.
63. Taubenberger, J. K., Reid, A. H., Lourens, R. M., Wang, R., Jin, G., & Fanning, T. G. (2005). Characterization of the 1918 influenza virus polymerase genes. *Nature, 437*(7060), 889–893.
64. Tian, J., Gong, H., Sheng, N., Zhou, X., Gulari, E., Gao, X., & Church, G. (2004). Accurate multiplex gene synthesis from programmable DNA microchips. *Nature, 432*(7020), 1050–1054.
65. *Tumpey, T. M., Basler, C. F., Aguilar, P. V., Zeng, H., Solorzano, A., Swayne, D. E., Cox, N. J., Katz, J. M., Taubenberger, J. K., Palese, P., & Garcia-Sastre, A. (2005). Characterization of the reconstructed 1918 Spanish influenza pandemic virus. *Science, 310*(5745), 77–80.
66. van Aken, J., & Hammond, E. (2003). Genetic engineering and biological weapons. *EMBO Reports, 4*(special issue), S57–S60.
67. van den Hoven, J. (1997). Computer ethics and moral methodology. *Metaphilosophy, 28*(3), 234–249.
68. Weckert, J., & Moor, J. (2006). The precautionary principle in nanotechnology. *International Journal of Applied Philosophy, 20*(2), 191–204.
69. Wheelis, M., & Dando, M. (2003). Back to bioweapons? *Bulletin of the Atomic Scientists, 59*(1), 41–46.
70. Zanders, J.-P. (2002). Introduction. *Minerva, 40*(1), 3–13.

Book Chapters

71. Briggle, A. (2005). Double effect and dual use. In C. Mitcham (Ed.), *Encyclopedia of science, technology, & ethics* (Vol. 2). Detroit, MI: Macmillan Reference USA.
72. King, J., & Strauss, H. (1990). The hazards of defensive biological warfare programs. In S. Wright (Ed.), *Preventing a biological arms race.* Cambridge, MA: MIT Press.
73. Lappe, M. (1990). Ethics in biological warfare research. In S. Wright (Ed.), *Preventing a biological arms race.* Cambridge, MA: MIT Press.
74. Marshall, J. M. (2004). Known stocks of the smallpox virus should be retained for research. In W. Dudley (Ed.), *Biological warfare: Opposing viewpoints.* Farmington Hills, MI: Greenhaven Press.
75. Reed, L., & Shulman, S. (2002). A perilous path to security? Weighing US "Biodefense" against qualitative proliferation. In S. Wright (Ed.), *Biological warfare and disarmament: New problems, new perspectives.* Lanham, MD: Rowman and Littlefield.

Books

76. Alibek, K., & Handelman, S. (1999). *Biohazard.* Broadway, NY: Dell Publishing.
77. Enemark, C. (2004). *Disease security in Northeast Asia: Biological weapons and natural plagues.* Strategic and Defence Studies Centre – Canberra Papers on Strategy and Defence No.156. Canberra: Australian National University.
78. Fenner, F., Henderson, D. A., Arita, I., Ježek, Z., & Ladnyi, I. D. (1988). *Smallpox and its eradication.* Geneva: World Health Organization.
79. Gross, M. L. (2006). *Bioethics and armed conflict: Moral dilemmas of medicine and war.* Cambridge, MA: MIT Press.
80. Henderson, D. A., Inglesby, T. V., & O'Toole, T. (Eds.) (2002). *Bioterrorism: Guidelines for medical and public health management.* Chicago, IL: JAMA and Archives Journals.
81. Kitcher, P. (2001). *Science, truth, and democracy.* New York: Oxford University Press.
82. Knezo, G. J. (2003). *Federal security controls on scientific and technical information: History and current controversy.* New York: Novinka Books.
83. Kolata, G. (2001, First published 1999). *Flu: The story of the great influenza pandemic of 1918 and the search for the virus that caused it.* London: Pan Books.
84. Layne, S. P., Beugelsdijk, T. J., Kumar, C., & Patel, N. (Eds.) (2001). *Fire power in the lab.* Washington, DC: Joseph Henry Press.
85. Miller, J., Engelberg, S., & Broad, W. (2001). Germs: *The ultimate weapon.* London: Simon & Schuster.
86. Miller, S. (2008). *Terrorism and counter-terrorism: Ethics and liberal democracy.* Oxford: Blackwell.
87. Moreno, J. D. (Ed.) (2003). *In the wake of terror.* Cambridge, MA: MIT Press.
88. Preston, R. (2003). *The demon in the freezer: The terrifying truth about the threat from bioterrorism.* London: Headline.
89. Regis, Ed (2000). *The biology of doom.* New York: First Owl Books.
90. Schweber, S. S. (2000). In the shadow of the bomb: *Bethe, Oppenheimer, and the moral responsibility of the scientist.* Princeton, NJ: Princeton University Press.
91. Stone, G. R. (2004). *Perilous times.* London: W. W. Norton.
92. Tucker, J. B. (2001). *Scourge: The once and future threat of smallpox.* New York: Atlantic Monthly Press.

National Academies Publications

93. Committee on Advances in Technology, the Prevention of Their Application to Next Generation Biowarfare Threats, National Research Council (2005). *An international perspective on advancing technologies and strategies for managing dual-use risks: Report of a workshop*. Washington, DC: National Academies Press.
94. Committee on Advances in Technology, the Prevention of Their Application to Next Generation Biowarfare Threats, National Research Council (2006). *Globalization, biosecurity, and the future of the life sciences*. Washington, DC: National Academies Press.
95. Committee on Balancing Scientific Openness, National Security, National Academy of Sciences, National Academy of Engineering, Institute of Medicine (2000). *Balancing scientific openness and national security controls at the nation's nuclear weapons laboratories*. Washington, DC: National Academies Press.
96. Committee on Genomics Databases for Bioterrorism Threat Agents, National Research Council (2004). *Seeking security: Pathogens, open access, and genome databases*. Washington, DC: National Academies Press.
97. Committee on Research Standards, Practices to Prevent the Destructive Application of Biotechnology, National Research Council (2004). *Biotechnology research in the age of terrorism*. Washington, DC: National Academies Press.
98. Committee on Science, Technology for Countering Terrorism, National Research Council (2002). *Making the nation safer: The role of science and technology in countering terrorism*. Washington, DC: National Academies Press.
99. Institute of Medicine, & National Research Council (1999). *Chemical and biological terrorism: Research and development to improve civilian medical response*. Washington, DC: National Academies Press.
100. National Academy of Sciences, National Academy of Engineering, Institute of Medicine (1982). *Scientific communication and national security: A report prepared by the panel on scientific communication and national security, committee on science, engineering, and public policy*. Washington, DC: National Academies Press.
101. National Research Council (2003). *Countering agricultural bioterrorism*. Washington, DC: National Academies Press.
102. National Research Council (2004). *Biotechnology research in an age of terrorism*. Washington, DC: National Academies of Sciences.

Websites

103. ABC Radio National (4 September 2005). BioWar – Background Briefing http://www.abc.net.au/rn/talks/bbing/stories/s1450124.htm Accessed 8 September 2005.
104. Alberts, B., Wulf, Wm. A., & Fineberg, H. (13 June 2003). Current visa restrictions interfere with US science and engineering contributions to important national needs, The National Academies http://www4.nationalacademies.org/news.nsf/isbn/s12132002?OpenDocument Accessed 10 June 2005.
105. American Medical Association (December 2004). Guidelines to prevent malevolent use of bio-medical research http://www.biosecuritycodes.org/docs/AMA%20Code.pdf Accessed 11 May 2006.
106. BBC News (19 April 2004). Caution urged over bio research http://newsvote.bbc.co.uk/mpapps/pagetools/print/news.bbc.co.uk/2/hi/health/3632349.stm Accessed 2 November 2004.

107. Biological and Toxin Weapons Convention (10 April 1972). Convention on the prohibition of the development, production and stockpiling of bacteriological (Biological) and toxin weapons and on their destruction http://www.opbw.org/convention/documents/btwctext.pdf Accessed 11 May 2006.
108. Central Intelligence Agency (3 November 2003). The darker bioweapons future http://www.fas.org/irp/cia/product/bw1103.pdf Accessed 11 May 2006.
109. Chamberlain, A. (August 2004). Select agent rules (updated), American Association for the Advancement of Science http://www.aaas.org/spp/post911/agents/ Accessed 3 December 2004.
110. CNN (6 June 2005). Feds: Science paper a terrorist's road map http://www.cnn.com/2005/US/06/06/milk.terror/index.html Accessed 17 July 2005.
111. Cohen, P. (18 January 2003). Recipes for bioterror: Censoring science, NewScientist.com http://www.newscientist.com/news/print.jsp?id=ns99993266 Accessed 2 November 2004.
112. Dayton, L. (27 October 2004). Scientists warn of 'Ethnic Weapons', *The Australian* http://www.theaustralian.news.com.au (generic URL) Accessed 27 October 2004.
113. Federation of American Scientists (November 2002). Draft recommendations for a code of conduct for biodefense programs http://www.fas.org/bwc/papers/code.pdf Accessed 30 May 2005.
114. Gammell, C. (25 October 2004). Biological weapons threat 'Greater Than Nuclear or Chemical', "PA" News, *The Scotsman* http://news.scotsman.com/ (generic URL) Accessed 27 October 2004.
115. Korch, G. (9 February 2004). Leading edge of biodefense: The national biodefense analysis and countermeasures center, Lecture to the Department of Defense Pest Management Workshop, Jacksonville Naval Air Station, Bioweapons and Biodefense Freedom of Information Fund http://www.cbwtransparency.org/archive/nbacc.pdf Accessed 26 May 2005.
116. MacKenzie, D. (29 October 2003). US develops lethal new viruses, *New Scientist* http://www.newscientist.com/hottopics/bioterrorism/bioterrorism.jsp?id=ns99994318 Accessed 2 November 2004.
117. McDonough, T. (23 February 2006). Just testing: Is the US government making anthrax bombs in Utah? *Salt Lake City Weekly* http://www.slweekly.com/editorial/2006/feat_2006-02-23.cfm Accessed 16 May 2006.
118. Nixdorff, K., & Bender, W. (2005). Biotechnology, ethics of research and potential military spin-off, *Bioterror and Bioweapons Control*, Bulletin 19, International Network of Engineers and Scientists Against Proliferation http://www.inesap.org/bulletin19/bul19art05.htm Accessed 28 November 2005.
119. Office of Biotechnology Activities (2005). National Institutes of Health, Frequently Asked Questions, National Science Advisory Board for Biosecurity http://www.biosecurityboard.gov/faq.asp Accessed 28 November 2005.
120. Petro, J. B. (2005). Intelligence support to the life science community: Mitigating threats from bioterrorism, Central Intelligence Agency http://www.cia.gov.csi/studies/vol48no3/article06.html Accessed 12 August 2005.
121. Roos, R. (6 November 2003). Scientists research antidotes to super mousepox virus, University of Minnesota http://www.cidrap.umn.edu/cidrap/content/bt/smallpox/news/nov0603mousepox.html Accessed 8 November 2003.
122. Ruppe, D. (3 February 2006). US army grenade patent changed, Global Security Newswire http://www.nti.org/d_newswire/issues/2006/1/31/76D99711-ECA8-4EA9-A2C2-278FCBD1159D.html Accessed 30 April 2006.
123. Select Agent Program (2005). The centers for disease control and prevention http://www.cdc.gov/od/sap/overview.htm Accessed 17 July 2005.
124. Shane, S. (27 June 2004). Bioterror fight may spawn new risks, Baltimore Sun http://www.baltimoresun.com/news/nationworld/bal-te.biodefense27jun27, 0,6098679.story Accessed 29 June 2004.
125. Sydney Morning Herald (12 July 2002). Scientists make polio virus using net recipe http://www.smh.com.au/articles/2002/07/12/1026185100298.htm Accessed 12 July 2002.

126. The Associated Press (10 September 2004). Panel: Germ data benefits exceed risk, *Newsday.com* http://www.newsday.com (generic URL) Accessed 13 September 2004.
127. The InterAcademy Panel on International Issues (7 November 2005). IAP statement on biosecurity http://www.royalsoc.ac.uk/displaypagedoc.asp?id=17463 Accessed 11 May 2006.
128. The Royal Society – Media Release (19 April 2004). Research should be vetted to prevent biological weapons development http://www.royalsoc.ac.uk/templates/press/releasedetails.cfm?file=521.txt Accessed 2 November 2004.
129. The Royal Society – Policy Statement (13 December 2004). Do no harm: Reducing the potential for the misuse of life science research http://www.royalsoc.ac.uk/document.asp?tip=0&id=2830 Accessed 2 June 2005.
130. The Royal Society – Press Release (1 December 2005). International scientists call to do no harm http://www.royalsoc.ac.uk/news.asp?id=3949&printer=1 Accessed 20 April 2006.
131. The Sunshine Project (October 2004). Mandate for failure: The state of institutional biosafety committees in an age of biological weapons research http://www.sunshine-project.org/biodefense/tspibc.pdf Accessed 10 June 2005.
132. The Sunshine Project (4 November 2004). High containment and other facilities of the US bio-defense program http://www.sunshine-project.org/biodefense/ Accessed 5 July 2005.
133. US Department of Health and Human Services – News Release (4 March 2004). HHS will lead government-wide effort to enhance biosecurity in "Dual Use" research http://www.hhs.gov/news/press/2004pres/20040304.html Accessed 27 May 2005.
134. US Department of Health and Human Services – Fact Sheet (21 July 2004). HHS fact sheet— project bioshield http://www.hhs.gov/news/press/2004pres/20040721b.html Accessed 27 June 2006.
135. US Department of Homeland Security (24 February 2005). Fact sheet: National biodefense analysis and countermeasures center http://www.dhs.gov/dhspublic/display?content=4336 Accessed 13 April 2005.
136. US Department of State (30 August 2005). Adherence to and compliance with arms control, nonproliferation, and disarmament agreements and commitments http://www.state.gov/t/vci/rls/rpt/51977.htm#chapter6 Accessed 22 December 2005.
137. Westphal, S. P. (17 July 2002). Ebola virus could be synthesised, *New Scientist* http://www.newscientist.com/article.ns?id=dn2555 Accessed 8 June 2005.
138. Wilson, J. (22 April 2004). Preventing test-tube terrorism, *The Guardian Review* http://education.guardian.co.uk/print/0,3858,4906769–110865,00.html Accessed 28 April 2004.
139. World Health Organization – Media Release (20 May 2005). Establishing smallpox vaccine reserve http://www.who.int/mediacentre/news/notes/2005/np_wha02/en/index.html Accessed 10 June 2005.
140. Wright, S. (November/December 2004). Taking biodefense too far, *Bulletin of the Atomic Scientists* http://www.thebulletin.org/issues/2004/nd04/nd04wright.html Accessed 1 November 2004.

Miscellaneous

141. Altman, L. (21 May 2005). WHO moves toward allowing smallpox gene experiment, *New York Times*, A6.
142. Anonymous (2001). US approves development of enhanced anthrax. *Arms Control Today*, 31(9), 26.
143. Australian Government (2005). Weapons of mass destruction: Australia's role in fighting proliferation. Canberra: Commonwealth of Australia.
144. Australian Safeguards, Non-Proliferation Office (2005). *Annual report* 2004–2005. Canberra: Commonwealth of Australia.

145. Borrie, J. (February 2005). The dual-use dilemma in life science research: The approach of the international committee of the Red Cross, United Nations Institute for Disarmament Research (UNIDIR) 'for comment' paper.
146. British Medical Association (1999). *Biotechnology, weapons and humanity*. Amsterdam: Harwood Academic Publishers.
147 British Medical Association (2004). *Biotechnology, weapons and humanity* II. London: British Medical Association.
148. Cohen, P. (April 2004). The individual and collective roles scientists can play in strengthening international treaties. The Royal Society – Policy Document.
149. Convention on the Prohibition of the Development (1975). Production and stockpiling of bacteriological (Biological) and toxin weapons and on their destruction (entered into force 26 March 1975) Available at: http://www.nuclearthreatinitiative.org/e_research/official_docs/inventory/pdfs/btwc. pdf.
150. Guidance for the Operation of Biological Resource Centres (BRCs) Part 3: Institutional biosecurity programme for BRCs – Discussion Paper (17 November 2005). Paris: Organisation for Economic Co-operation and Development.
151. McLeish, C. A. (2003). Reactions to self-censorship, Working Paper given at the 20th Pugwash workshop study group on the implementation of the CBW conventions: *The BWC intersessional process towards the sixth review conference and beyond*. Geneva, Switzerland, 8–9 November 2003.
152. Miller, J., Engelberg, S., & Broad, W. (4 September 2001). US germ warfare research pushes treaty limits, *New York Times*, A1.
153. Mordini, E. (April 2004). Conclusions of the international conference on ethical implications of research into the prevention of bioterrorism, Bioethical implications of globalisation processes – Policy Paper No 1, Conference organised by European Commission, DG Research, Directorate E: Biotechnology, Agriculture and Food.
154. Pearson, G. S., & Dando, M. R. (Eds.) (September 2004). Towards a life sciences code: Countering the threats from biological weapons, *Strengthening the biological weapons convention – briefing paper No 13 (Second Series)*. Bradford: Department of Peace Studies, University of Bradford.
155. Pearson, G. S., & Dando, M. R. (Eds.) (November 2004). A code of conduct for the life sciences: A practical approach, *Strengthening the biological weapons convention – briefing paper No 15 (Second Series)*. Bradford: Department of Peace Studies, University of Bradford.
156. Pollack, A. (12 July 2002). Scientists create a live polio virus. *New York Times*, A1.
157. Smith, D. (29 December 2003). Call to stop deadly viruses getting into wrong hands. *Sydney Morning Herald*, 5.
158. Stockholm International Peace Research Institute – Fact Sheet (November 2001). Biotechnology and the future of the biological and toxin weapons convention.
159. The Royal Society– Policy Document (2005). The role of codes of conduct in preventing the misuse of scientific research.
160. Tucker, J. B., & Zilinskas, R. A. (Eds.) (2002). The 1971 smallpox epidemic in Aralsk, Kazakhstan, and the Soviet Biological Warfare Program, *Center for Nonproliferation Studies, Occasional Paper No. 9*, Monterey. California: Monterey Institute of International Studies.
161. US Department of Health and Human Services (January 2003). *NIAID biodefense research agenda for category B and C priority pathogens*. NIH Publication No. 03–5313.
162. US Department of Health and Human Services (4 March 2004). National Science Advisory Board for Biosecurity Charter.
163. US Office of the Under Secretary of Defense for Acquisition and Technology (1998). Militarily Critical Technologies List (MCTL), Part 2: Weapons of mass destruction technologies.
164. WHO Advisory Committee on Variola Virus Research (2006). Report of the seventh meeting, 10–11 November 2005. Geneva: WHO.

Index

A

Aerosolization
  and mass-vaccination, 25
  pathogen, 11, 33
African National Congress (ANC), 17
American Medical Association (AMA), 47, 59
Anthrax
  drug resistance, 21
  gene sequencing, 26
Arctic permafrost, 29, 33

B

Bacillus anthracis. *See* Anthrax
Biological weaponization, 24–26
Biological Weapons Convention (BWC), 12–13, 17, 18, 53
Biopreparat program, 39
BioShield project, 11, 62
Biotechnology Research in an Age of Terrorism, 47

C

Camelpox virus, 21, 23
Chemical Weapons Convention (CWC), 13
Chimera, smallpox and ebola, 22, 23
Cidofovir, efficacy test, 21
Clear Vision project, 24
Cowpox virus, 20

D

Diagnosis/detection disruption, 23–24
Drug-resistance, 20–21
Dual-use dilemma
  definition, 1
  report, aim and scope of, 2–3
  research context, 9–12
Dual-use research
  and academic freedom, 40–42
  benefits and burdens, analysis, 3–4
  dynamic analysis, 4, 32
  experiments of concern, 17–30
  moral permissibility, 4, 14, 55
  physical safety and security, 50
  private sectors, role of, 48
  regulation options, 5–8
  scientific progress, 43, 44
  smallpox debate, 4

E

Ectromelia virus. *See* Mousepox virus
Experimental research
  biological weaponization, 24–26
  creating drug resistance, 20–21
  diagnosis/detection disruption, 23–24
  enhancing pathogen virulence, 21–22
  increasing pathogen tranmissibility, 22–23
  micro-organism genetic sequencing, 26–27
  past pathogen revival, 28–29
  pathogen host range alteration, 23
  pathogenic micro-organism synthesis, 27
  rendering vaccines ineffective, 19–20
  smallpox virus (variola), 28

F

Fenner, F., 33
Fine-grained ethical analyses, 31, 32
"Fink report." *See* Biotechnology Research in an Age of Terrorism
Flu resurrection study, 47

G
Governmental control, in dual-use research, 57–59, 61, 62

H
H1N1 influenza virus, gene sequencing, 26, 27
H5N1 avian influenza, 22
Henderson, D. A., 33
Human health and ethics of research, 32

I
Impermissible research, 14
Independent authority, in dual-use research
- advantages, 60
- disadvantages, 60, 61
- role of, 59

Institutional biosafety committees (IBCs), 55, 59–61
Institutional control, in dual-use research
- advantages, 56
- disadvantages, 56, 57
- permissibility of research, 55

Intellectual inquiry freedom, 40, 41

J
Jefferson Project, 20

K
Kolata, G., 29
Korch, G., 25

L
Licensing of dual use research, 50, 51

M
Microencapsulation of pathogen, 24
Micro-organism genetic sequencing, 26–27
Monkeypox virus, genetic mutation, 33
Moral permissibility, dual-use research, 14, 55
Mousepox virus, 12, 19, 21
Myxoma virus, mutation, 23

N
National Biodefense Analysis and Countermeasures Center (NBACC), 25
National Institute of Allergy and Infectious Diseases (NIAID), 20
National Research Council (NRC), 18, 19, 47
National Science Advisory Board for Biosecurity (NSABB), 26–27

P
Past pathogen revival, 28–29
Pathogenic micro-organism synthesis, 27
Pathogens
- aerosolisation, 11, 33
- elimination, 33
- and host range, 23
- host range alteration, 23
- transmission, 22, 23, 34
- virulence, 21, 22, 34
- and weaponisation, 36–38

Pathogen tranmissibility increase, 22–23
Pathogen virulence enhancement, 21–22
Poliomyelitis (polio) virus, genome synthesis, 27
Proceedings of the National Academy of Sciences (PNAS), 22
Project Clear Vision, 24
Project Jefferson, 20

R
Ramshaw, I. A., 46
Research
- biological weapons development, 17–19
- censorship, 52
- experiments of concern, 19–20
- information dissemination, 14
- moral permissibility, 14, 55

S
Scientist's autonomy
- advantages, 53
- disadvantages, 53–55

Severe acute respiratory syndrome (SARS), 23
Smallpox-Ebola chimera, 22, 23
Smallpox eradication campaign, 33, 34
Smallpox virus, antiviral drugs development, 28
SPICE protein, 22
Statement on Scientific Publication and Security, 44–48

T
Tuberculosis (TB), virulent from, 22

V
Vaccine-resistant mousepox, 19
Vaccines, rendering ineffective, 19–20
Variola Advisory Committee (VAC), 28, 34
Variola major virus, 28. *See also* Smallpox virus

W
Weapons of mass destruction (WMD), 9, 17, 18
Weckert, J., 35
Wimmer, E., 27
World Health Assembly (WHA), 28, 34
Wright, S., 35

GPSR Compliance

The European Union's (EU) General Product Safety Regulation (GPSR) is a set of rules that requires consumer products to be safe and our obligations to ensure this.

If you have any concerns about our products, you can contact us on ProductSafety@springernature.com

In case Publisher is established outside the EU, the EU authorized representative is:

Springer Nature Customer Service Center GmbH
Europaplatz 3
69115 Heidelberg, Germany

Batch number: 10830943

Printed by Printforce, the Netherlands